A WILDER LIFE
Essays from Home

A Wilder Life

ESSAYS FROM HOME

Ken Wright

KIVAKÍ
PRESS

Some of the essays in this book have appeared in somewhat different
form in the following publications: *Backpacker, Sierra, San Juan Almanac,
Winter Park Manifest, The Sunday Camera Magazine, The Durango Herald,
Images, Walkabout, and Rocky Mountain News.*

Kivakí Press, Inc.
585 East 31st Street
Durango, CO 81301

Publisher's Cataloging in Publication

Wright, Ken, 1960-
    A wilder life : essays from home / Ken Wright.
    p. cm.
    ISBN: 1-882308-15-8

    I. Title.

PS3573.K46W55 1995   814'.54
                                  QBI95-20085

Preassigned Library of Congress Catalog Card Number: 95-60389

Kivakí Press offers special discounts for bulk purchases of books for promo-
tions, premiums, and non-profit or educational organizations.
Arrangements can be made for subsidiary publication of special editions or
book excerpts. Address inquiries to:

Kivakí Press Special Sales
585 East 31st Street • Durango, CO 81301
303-385-1767 tel • 303-385-1974 fax
kivaki@frontier.net (Internet)

Book design and illustration by Olive Charles
Text set in Garamond
First Edition
First Printing 1995
Printed in the United States of America on recycled paper
1 3 5 7 9 10 8 6 4 2

We live at a time when we chafe at the restraints that our minds and our customs and our glimpses of the future have imposed upon us. We cannot imagine any longer a situation that does not call for rules, often new rules. We have created a culture where even the idea of wilderness is encapsulated within regulations, permits, and strict decorum. Perhaps there is nothing to be done about these facts. Still, deeper impulses linger and whisper to us acts of rebellion....

—Charles Bowden, *The Secret Forest*

This book is dedicated to:

my mother,
who taught me how to treat people,
my father,
who taught me how to treat the land,
Sarah,
who taught me how to treat myself,
and Webb,
who taught me how to treat the future.

🌿 🌿 🌿

CONTENTS

🌿 🌿 🌿

Introduction: The Weed People 1
1. Swamp Waters. 5
2. Ken's Shortcuts 23
3. Dark Canyon: A Walk On The New Frontier 25
4. Good News. 38
5. Some Thoughts On Growth. 40
6. Widening Trails 43
7. The Great Kiva 57
8. Drinking and Driving 58
9. My Greatest Fear 65
10. The Valley. 66
11. A Modest Proposal for the Interior West. 71
12. What Matters 87
13. Wild Child: Why The Environmental Movement
    Needs Parents and Children 88
14. Sarah. 94
15. Ode To Edward Abbey, Part 1: A Death In the Family . 96
16. Ode To Edward Abbey, Part 2: Abbey Lives! 101
17. My Life In A Nutshell 107
18. Let Us Now Praise The Colorado Squawfish 108
19. Thoughts On A Child's First River Trip 111
20. On Hunting 116
21. Reclaiming The New Terra Incognita 119
22. Why Free-Market Environmentalism Will Fail . . . 123
23. A Married Man 124
24. Time To Get Out. 130
25. In The Forest. 132
26. Final Thoughts: In Bed One Night. 161

🌿 🌿 🌿

INTRODUCTION: THE WEED PEOPLE

> To be a philosopher is not merely to have subtle thoughts, nor even to found a school, but to so love wisdom as to live according to its dictates, a life of simplicity, independence, magnanimity, and trust. It is to solve some of life's problems, not only theoretically, but practically.
>
> —Henry Thoreau, *Walden*

I AM DRAWN TO ODD PEOPLE.

It doesn't bother me, but I find myself constantly attracted to and surrounded by an odd assemblage of friends—or an assemblage of odd friends—who don't float the normal currents of society. You don't see people like them on T.V. Politicians don't seek their support. Advertising campaigns don't target them. It seems these friends of mine are willing to put style over fashion, quality over quantity, place over practicality, risk over haven, richness over riches, friendships over acquaintances, knowledge over college, ritual over routine, spirituality over religion.

I've been thinking about this lately, and I've come to some

conclusions about this unique and distinctive collection of friends I've acquired over the years. I have uncovered a common thread weaving together myself and my uncommon compatriots. I notice that my closest companions are lovers—lovers of places, lovers of life, lovers of freedom. They are lovers of rivers and rocks and trees. They are lovers of adventures and chores and challenges. They are lovers of their spouses, their friends, their kin, and their communities.

But they're more than just lovers.

I notice that the people I find most alluring are also fighters—defenders of the things and people they love, defenders of the ideals that fuel and guide their loves. Lovers and fighters: the two personalities inseparable.

These friends and I are bonded by a vision of love that directs us like a compass bearing guiding the walk of life: we see love not as an object, but as an action; we use the word "love" not as a noun, but as a verb. This loving is uncompromising and demanding, fueled by appreciation, defense, and celebration. Appreciating by seeing the loved for what it is, not what we wish it were; defending by standing by the loved one's unique self and its right to blossom as its nature dictates; and celebrating our love by fulfilling our psychological and spiritual need for awe, magic, wonder, rituals, and traditions that fuel honest appreciation and unwavering defense.

This is nothing new. These people I describe inhabit the modern world by circumstance, but they live in it with the nature-borne spirit of joy, independence, and need for wild landscapes that humans lived with for hundreds of millennia, the same spirit that today's industrial, technical, machine-run and money-built, urban, paranoid, land-detached, and economic-growth-driven society is burying.

These people are weeds sprouting through that pavement. They live in ecosystems, not economic systems; they are guided by conscience, not dogma; they get their news from the out-of-doors, not out of news reports; they seek real experience, not virtual reality.

This is why my friends are odd.

And what should we label these odd friends, and others like them? They are not easily categorized, summarized, labeled or stereotyped. They are diverse and varied and changing. No word seems to encompass them although many words apply at different times, from "environmentalists" to "rednecks" to "Generation X'ers."

So I made up my own label: I call them *wilders*. I chose wilders because these people don't just believe in a wilder way of seeing the world and of living, they do it. An "ist" believes, but an "er" does.

Wilders.

And to live with their spirit in its many manifest forms is to live the life of a wilder, a wilder life.

What do I mean by a wilder life? Answering that is the purpose of this book. Still, this book isn't so much an explanation as it is a collection of ideas and stories written from a wilder's perspective. Rather than trying to thoroughly and rationally explain anything, in these pages I just look and think and speak as honestly as I can from one wilder's ideals. Show, don't tell—a rule of living as well as writing. You be the judge of the result.

Of course, I couldn't have undertaken this endeavor alone. I would like to thank the friends and fellow writers who have offered support, inspiration, advice, and criticism when it was needed. Most notably Mark Spitzer, Kevin Burns, Gregory and Julie Moore, John and Mary Drigot, Bob and Eve Meyer, Tommy Mountain, Lisa Lenard, Pam Houston, Chad Learch, Chris Goold, and Greg Cumberford of Kivakí Press, who was brave enough and fool enough to take a risk on this book.

I also can't forget the many writers and musicians who have molded me, who are the elders leading the way. As philosopher Thomas Berry said: "Beyond the country's political and economic needs, and possibly a prior condition for any sustainable political structure or functional economy, is the need for a mystique of the land such as is supplied by the nature poets, essayists, and artists." For their words and inspiration, I am especially indebted to Edward Abbey, Terry Tempest Williams, Dave Foreman, Charles Bowden, David Petersen, David Seals, Sigurd Olson, Wallace Stegner, H.D. Thoreau, Jonathan Edwards, Chuck Pyle, Pete Townshend, and Bruce Cockburn.

Bless you all.

1

Swamp Waters

I DON'T USE THE WORD "BLISSFUL" VERY OFTEN, BUT IT'S THE ONLY WORD THAT FITS WHAT I FEEL AS OUR MOKORO glides through the grasses and over the waters of these swamps.

Here in southern Africa, the continent's third largest river, the Okavango, flows off the mountains of Angola. For 600 miles the river meanders across the Kalahari Desert in Botswana before fanning out to create a puddle of life in the midst of the immense Kalahari Plain.

This is the Okavango Delta, a 10,000-square-mile area of swamps and islands, and the largest inland delta on Earth. Where the Okavango River dies an evaporating death—only 3 percent of the water that enters the swamps flows out—life is given to one of the largest, most diverse and still-thriving wildernesses on the planet.

And this is where my wife, Sarah, and I have chosen to honeymoon. As I see it, the experience is a sort of wilderness rite-of-passage to enter the wilderness of marriage. The wilderness of marriage? As I look at the greenery around me, the water, the treed horizon, it all looks familiar as wild country—I have built my life around living and working in wild country—yet it is all alien; the trees and flattened terrain and expanses of clear water are all new to this native North American.

Like marriage. My wife sits behind me in this narrow craft. If I were to turn around and look, I would see a pretty redhead peering about, taking in the scene and scenery. She would look familiar, but something would be new to me. She is my *wife*. Sitting there is a person I will spend my whole life with, make all my decisions with, grow a family with. New terrain, a personal wilderness spread out before me, beckoning, intimidating. I don't look.

If I did look, though, I would also see another person, for we are not alone in this hand-carved canoe. Obeilwe Repiusa stands in the back of the boat with a 15-foot pole he uses to move us through the swamp waters. We've been calling him "O.B." at his suggestion, I assume to preserve the integrity of his true name, which we can only mumble painfully.

O.B. is about my height, five-nine, thin but finely toned, and the color of rich soil. Despite the well-worn formal attire he wears, he merges with this place, whether poling from the rear of his hand-made craft or ambling agilely through the thick, scrubby bush. These swamps are his home, and it shows as he moves the mokoro through the thick grasses via pathways worn by generations of such passings. This is the highway system of the indigenous swamp dwellers, and only a person of the swamps is able to navigate it. As if to prove this, O.B. guides us to a seemingly impregnable wall of reeds, then turns into a narrow green alley.

For Sarah and me, this is our third day in the swamps and our first in the mokoro. But the adventuring began days before this, the commercial safari portion of our Okavango adventure. To get here from the edge of the delta, at the northwestern Botswana frontier town of Maun, involved hitchhiking, four-wheel-drive trucks, and a six-seater airplane. In those days we passed from drinking South African wine in an Okavango resort to drinking beer with Tswana villagers on a delta island.

"Crocodiles eat only when they're hungry," a hand-painted sign over the island bar said, "but they kill anytime." The Tswanans I talked to, most of them mokoro guides, at that wooden-plank watering hole confirmed the wisdom, nodding and grinning. I was a little skeptical of the drama. Since Sarah and I are commercial river guides ourselves, I understand a guides' needs to inflate, just a little, the aura of their home turf. That way customers feel they get their money's worth in danger even on the mellowest trips. And it helps the tips. And it's good for the guides' egos, also generally inflated. Back home we might shoot for getting customers to tell their friends, "My God, we had a smooth run down those rapids. You should've heard the stories...." Here the goal might be, "The crocs were everywhere, but our guide was bloody fine...." So while Sarah retrieved her underwear from the baboons that had broken into our tent again, I bought the Tswanans a round of beers and told them horror stories about the rapids on Colorado's Arkansas River.

But this isn't Colorado. I haven't seen any crocs, but the rough and hard feel of the mokoro reminds me of where we are. The self-bailing, fat-tubed rafts we use on Colorado's rivers are cushy sofas compared with the hand-carved wooden mokoros of the Okavango. Once in the swamps, though, a mokoro is the only way to travel, literally. It's not a bad way, either. As I lean on my pack, the low, narrow craft hisses along, displacing little water, parting the reeds with little effort. It is part of this landscape, crafted from swamp-grown wood. O.B. stops at one point, walks us onto to an island, and proudly shows us the stump of the tree that bore this canoe. It took him three months, he says, to carve the fifteen-foot-or-so length with an axe and a hatchet. A mokoro like this will last nine or ten years. And when its floating days are done, the ends are cut off. The body of the canoe becomes a camp bench and the severed ends become chairs, until they return to the soil.

We float through a marsh of green and brown grasses, water lilies and purple flowers. From the front of the mokoro, the view is like a movie unfolding. Each parting of the grasses yields a new scene, a new act in the story of the delta, a new subplot for the wild residents. At least that's how it seems to me, as I sit and watch, eagerly relaxed.

This is blissful.

✦ ✦ ✦ ✦

Several hours pass quietly. Despite the busy traveling involved in getting here, the recent insanity of the wedding, the packing, the two days of flying, Sarah and I adjust quickly to the calm of our floating. We are silent, and O.B. poles us along easily, slowly, steadily.

Then we hit shore. Suddenly and without warning, O. B. turns and pushes the mokoro onto solid ground under a grove of tall trees. He steps around us to shore while Sarah and I wobble our way out of the shaky craft. As we feel for our land legs, O.B. trots to the edge of a meadow a short distance from shore. He peers ahead from the thicket, then silently beckons us with a wave. In the meadow, a waterbuck watches as Sarah and I join O.B., then it leaps away, the white circle around its brown rump following the spiral of his horns. A significant event: This is a new mammal for us. Sarah and I take the sight as a good omen, as though we have been welcomed to this place, and a meeting of our eyes seals the experience.

We set up camp under the trees, nibble some lunch, then O.B. leads us on our first foot-safari into the bush. Sarah and I wander along in ultra-light hiking boots, shorts, T-shirts and hats while O.B. walks in dress shoes, polyester slacks, a button-down collared dress shirt, and a faded "Camp Okavango" visor. We can barely keep up with him. We crash and crunch through the bush while he makes no sound in his battered penny-loafers.

Wildebeest, impala, tsesebe, kudu, baboons, and warthogs. Today six new animals enter our lives. We see them hidden in the dense foliage, running in long umber lines of fifteen and twenty, as flashes bouncing through the woods, and standing in compact and social little groups, confident that this is their place as we watch respectfully from the woodline.

And what do I feel encountering these large mammals for the first time? Awe? Delight? Mystery? Fear? All the above, perhaps. But the most surprising sensation is one of familiarity. A "don't I know you from somewhere?" sense. As I gaze from the cover of trees out over plains and meadows where these large and wild herds of foreign (to me) creatures run and graze and generally lounge, I sense a surge of genetic memory. We evolved together, these creatures or some like them and my ancestors, living with each other, off each other, needing each other. I may be used to stalking my dinner down the corridors of City Market, but now as I watch still and silent from the forest's edge—as humans have for most of their evolutionary lives, and as we will again—I am reminded that the millennia dividing our histories is only a thin veneer.

Those are the thoughts I think as O.B. explains something in a whisper to Sarah. She follows his pointed finger with her eyes. I hang with my thoughts, which take a surprising turn. I think of Sarah and me. I look out over this wild scene and see Sarah, also looking out, face placid and pensive under her floppy canvas hat. And I find myself gazing from the edge of our marriage, scouting ahead, searching for a route through the life we launch here. What does our marriage have to do with big ungulates lolling on the African plain? Again, something is familiar here: I see a vision of our life and I sense that what guides us and binds us is, perhaps, not so much our similar lifestyle choices as it is some genetic compass whose bearings we share.

Sarah and me: Also like our ancestors did for hundreds of

millennia, we live close to the land, in remote places, near wild country, often in tents on the ground. In our own modern way we are still migratory, building our careers around seeking snow in winter and mountain rivers in the summer. In the spring, we follow the warmth as it moves north, and in the fall we nest. While we live this way now, we both look to end our migrating in coming years and find a place to stop and take a stand, to immerse ourselves in the human needs for kin, community, and a landscape we know intimately.

Our relationship is built on these desires, these drives. These are our highest values in life, higher than our incomes and job security and taking the easiest path in life, for there are easier paths we could take. And what are these desires and drives? Aren't these wild sensibilities? Aren't these variations on a wild theme?

Under an infinite African sky the herd grazes along slowly, unhurried. O.B. smiles at us, and we move on.

♦ ♦ ♦ ♦

At sunset the sky is several shades of periwinkle and aquamarine, checkerboarded with glossy pink clouds. The evening air is cool but dry. It's June, the middle of winter, and no rain is predicted for the next six months. Since it is the African winter here, the sun sets around 6 p.m. and it doesn't reappear until 7 a.m. Normally in this month Sarah and I are in our summer mode, living on the river and in close alignment with the cycles of the sun. So the brevity of this African day confuses our bodies.

Still, the shortened day doesn't seem to matter tonight as we sit around our dinner fire. I am tired, but feel invigorated and alive. There is a brilliance to the moment: An evening zephyr brushes my left cheek, the flat sun warms the right. My worn Red Sox cap sits snugly on my head. I smell thin smoke and chicken soup. I feel remarkably *present,* a heightened awareness that is normally lost in the static of the day. Retrieving this awareness is why I go to the wilderness. Just think: If we could keep this edge on our awareness all our lives, day after routine day at work and on commutes and in our dreary daily dealings rather than letting these things dull that edge, then what richly alive lives we could carve. Our lives would pulse and throb with the distinct details of each unique moment. But it seems constant contact with the grindstone wears down that edge of awareness.

And that's a selfish reason we need wilderness, because venturing there hones an edge back onto our momentary, immediate, quiet-yet-quick animal-sense awareness, the awareness that transforms moments into eternities, that imbues the world with magic, that surprises us with the power of the senses.

But I ask myself again: Why can't we live each day with that wilderness-sharpened awareness? Not merely as a metaphor, but as a style, a reality, an awareness that sees all the world as magic, and ourselves as powerful and involved actors despite routine and concrete demands and obligations? Why can't we just be camping, day by day? Hunting and gathering in the modern world? I think about this a few minutes, then decide, shit, I don't know. Maybe it's a law of nature that our industrial lifestyles wear that edge off our awareness. Or, maybe it just doesn't occur to us to not let that lifestyle grind the edge off.

Well, it occurs to me now, so I breathe in the scene: a campfire in Africa, my new wife, a new life, the same old sun rolling off the edge of the same old planet toward a new night and a new day. Maybe I can commit to a new mission to go along with all this newness: From now on there will be no break between my wilderness adventures and my day-to-day life, between the wild places I visit and my work, my chores, and my family life. My house will be an extension of my tent and not the other way around. I will head out on a life-long walk, and the ground I cover daily will be the edge on which I will hone my awareness of the sublime in each day.

A nice idea. It feels good and right. I smile at Sarah. O.B. watches, perplexed, as Sarah and I face west and wave good-bye to the sun, a custom of ours that confirms my suspicion that it's a good thing we've married each other.

Night falls, but we cannot consider sleep yet. It's all so new, and we—Sarah and I, at least—don't want this day to end. What does O.B. think? Is this routine to him, just another night in the Okavango? Does he wish he could get "Cheers" on T.V.? He sits quietly, weaving a fishing net from cord salvaged from automobile tires. He drives a metal rod into the ground, then ties a piece of cord off on the rod. He wraps another piece of cord around the tied one and draws toward him, stripping off the excess rubber chunks. He saves the cleaned cordage to be woven into a net later. I watch closely. He smiles when our eyes meet, never breaking his unhurried and deliberate rhythm. Hunting and gathering in the modern world.

The three of us sit in a dark circle illuminated by a tall fire that dances between us. Night birds whistle, frogs grunt and bell frogs keep time with a ringing tone like wood on hollow wood. "Life" is the word for this place. Everywhere is life. Every corner of the landscape is full of rich, abundant movement and the sounds of life. There is here a diversity of living things beyond any of my previous experience—in the trees, the trees themselves, in the bush, in the water, in the air. It's the antithesis of the sparse desert land I am used to. The Okavango literally breathes and writhes. The swamp glows in the light of the full moon like a frosted meadow.

In the morning, after a quick breakfast of stale cereal and Ricoffy, an instant coffee and chicory blend that passes for coffee in these parts, the camp comes down. We load, and O.B. and his mokoro take us out into the swamp where we are caressed by a cloudless sky and cool breezes. We are immersed in the living greens that are reeds, grasses and pure waters—so pure we dip our cups and drink when we are thirsty. The ability to do this, to drink from our home's rivers and lakes, it seems to me is such a natural-born right as a resident of this planet that I have never understood why we let anyone anywhere take that away. When did polluting become a right and pure wild water a luxury we must pay for?

"Some people are nice," O.B. said once when we were discussing the customers he has had here. "Some not nice." Sums up a lot, I think, as I reach over the side of the mokoro for another cup of water. As I do this, I see suspended bits of Styrofoam float by and, clearly visible in the very clear water, a beer can mired in the mud. Not here, please, I beg. Not here, too. But, as O.B. says, *some not nice.* I think back to the shanty island bar where I sipped Castle Lagers with my new Tswana friends, my fellow guides. Before Sarah returned from retrieving her underalls, two Brits joined us. Suddenly my friends turned their conversations inward and back to their native tongue. The two English gentlemen talked to me, in our native tongue.

More Castles were ordered. I asked them who they were.

"Two African explorers off to conquer and colonize the last untamed wild lands for the empire!" one answered. They laughed.

After a few minutes, the beers still hadn't arrived even though there were only a few of us at the bar. The other Brit yelled at the Tswana bartender, then turned back to me.

"Shoot one and the rest will listen, I say," he mumbled.

Yes. Quite so, I thought, regretfully unarmed.

Later that night, Sarah and I and the mokoro guides got a little drunk, and moved our festivities outside to a big fire surrounded by dead-mokoro benches. Our thin language link didn't keep us from laughing and joking. Then the South Africans came. A couple from Johannesburg—"Jo-burg"—joined the swinging ring around the fire. It didn't take long for them to attach themselves to us, the only other whites present. It didn't take long for Otto to tell me why he wasn't laughing like rest of us.

"We live with them, so we're sick of their trying to act white all the time."

Thanks for clearing that up, Otto.

"Look at them," he mumbled on. "You can take them out of the bush, but you can't take the bush out of them."

Hallelujah. May that be the epitaph on my tombstone.

There was a sign of hope that night, though. Later on two young white South African girls joined us. They drank beer and laughed with the Tswanans. They chatted with Otto and his overdressed wife, then moved on to me and Sarah. In the course of our conversation one glanced at the other Jo-burgers. "Fucking fascists" she called them. Bless you, child, I thought. Bless the hope for the next generation.

I sit back up and the Styrofoam floats away. There's not much of this civilized jetsam here, but any amount is too much. This is perhaps the largest area of true wilderness left on the planet, the only place where even the sky is sacred, spared the dividing ribbons of jet contrails, so why can't we just leave this alone? I think of my own country, of the great parks of Colorado and the wild rivers of the Rockies; I think of what it used to be like, with great herds of bison and packs of wolves. Was it like this place, rich in the diversity of life and pure air and water? Where wild people led wild lives? What happened? What the hell happened? Why did we *let* it happen?

I think I'd rather not think about that right now. Not here. Not on my honeymoon. Not when we're gliding silently up to an unsuspecting reedbuck, standing waist-deep in the water in front of us. None of us moves except Sarah. The click of her camera sends him bounding through the marsh, throwing water high in the air.

We need this, big wilderness like this, to remind us of what was, to tell us what should be, and to guide us in what we have to do. There is always hope in the next generation. I hope.

✦ ✦ ✦ ✦

We land on another island. It's hard for me and Sarah to predict how big each island is, since to our untrained eyes all this terrain looks like mostly terra firma laced with moats. Most of what looks like solid ground is, of course, merely dense plants camouflaging water. Swamp. Any island where we can see solid land under trees could be either the size of a suburban back yard or Delaware—we can't tell without getting out and walking. But fortunately we have O.B., and O.B. tells us this is a good island, a big island. We leave our gear in the mokoro, the mokoro unlocked, and we amble out through trees and across open grassland under a big sky.

We walk along briskly, O.B. stopping every now and then to look at the ground, at a shrub, at that sky. I'm not sure at what or for what he looks, but he occasionally borrows my binoculars, and every now and then he points out birds in flight or roosting in trees—bee eaters, battleurs, pied kingfishers—then we walk on. The air smells acrid and dry.

In a thicket of trees and shrubs, O.B. stops. He turns slowly toward us and points upward. I don't see anything. Sarah looks at me, eyebrows furrowed. She doesn't see anything either, just the darkness between the leaves in the wall of trees O.B. points to. He still points, wide eyes speaking for his silent mouth—*Look! Look!* I follow his finger again, but see nothing but trees and branches and foliage and the space between the leaves and...then I look up, to the tops of the branches, to the space between the crowns of the trees, and I see a head, a high-top-shoe-shaped head with two thick antennae rising behind little ears, a head above the trees, nibbling the tender high branches. A giraffe. Three of them: once I identify one head, two more become clear, even obvious. I've seen some weird and wonderful things in life, but this wins, no contest.

O.B. doesn't let us stare dumbfounded for long. He seems still on a mission. A quick wave and he trots off to the right of the giraffe-headed trees.

We run for ten minutes, then O.B. stops. He jumps on the fat reclining body of a fallen tree and gazes outward. We join him and look over another grassy plain, a huge meadow like a great lake over which we can't see a far shore. And there, only some forty or fifty yards away, is a family of giraffes. Two big adults with dark patterns of amoebae-shaped spots. Four smaller giraffes follow the bigger ones, heads swinging forward and back with each step like slow, giant pigeons. Then they all run, then stop, then run. We watch them move away for a half an hour, Sarah and I arm in arm and O.B. just watching.

✦ ✦ ✦ ✦

Another evening, another camp, another campfire, another tremendous big-sky sunset. To the east it is night-dark and to the west the colors of the spectrum are laid out in sheets. It is so flat here on the Kalahari Plain that as the rays of the sun bend unbroken through the atmosphere the trees reflect a rosy alpenglow like Colorado's mountains. Sarah and I wave to the west. "Good-bye, sun," she says.

As the stars emerge I am again in awe of the appearance of the Southern Cross. It has been a life dream of mine to see that. The rest of the stars are in an unfamiliar arrangement; it's our galaxy, I know, but as if seen from an alien planet. I am free to arrange my own constellations: there's a can of beer, there's a banjo, there's Carl Yastrzemski. On that note, I head to bed.

✦ ✦ ✦ ✦

Morning. A quick breakfast, toss some snacks in a day pack, and it's off into the bush again. O.B. is the consummate guide today, identifying tracks in piles of dirt, discovering birds and animals in thickets, and leading us up humongous ant hills, the only relief to this flattened topography, so we can scan the landscape.

Sarah is becoming quite the African guide herself, pointing out animals and birds almost as quickly as the hawk-eyed

O.B. (Sarah has 20-15 vision, another good reason to marry her.) She borrows O.B.'s guidebooks to search for information on a white-cowled Fish eagle that hangs in the air over us. As I look at her under that floppy cotton hat and hand on O.B.'s shoulder as he helps her find something in the bird book, I thank my good fortune to be married to this person. There's a whole planetful of places to walk, and I want to walk them all with her.

Which brings up another thought. I think about women. Women and wilderness. We men need both, I believe, together. Even though I was weaned on the tales of the pathfinders and explorers and John Wayne and Johnny Quest, and even though I cut my wilderness teeth with my father and his close band of outdoorsmen compatriots, I never really understood the traditional male explorers' great adventuring and discovering with only other men. Great groups of men. Men in action. Men leaving the women behind to take care of home and hearth and raise the kids.

I hear these tales and ask, where are the women? Women should have been there and should be there, not because they have "the right" to be there, not because of some "equality" rationale, but because women offer different perspective and spirit. I want women there because it's *better* that way. Really, who wants to spend days, weeks, months, even years in the wilderness with a bunch of men? And really, when you think about it, why is that considered normal?

Just wondering.

Kudo, impala, lechwe. Red horn bill, Fish eagle, battleur. The wonders keep coming.

✦ ✦ ✦ ✦

A new camp, again opened with a good omen as impala mill about, reluctantly yielding this fine, large, dirt clearing in the trees. This island is another big island, O.B. assures us. Chief's Island. We lounge around another fire, nature's T.V. (never a rerun). Sarah soon retires. She says goodnight to us, leaving me alone with this man from another world.

We speak little, but enjoy each other's company. Speech is a remarkably small part of communication. Our attempts to learn about each other come out in simple sentences—sentences spoken in English, of course, reflecting a hierarchical situation of the world that seems like a rumor here, an impossible tale. Reality is different. In the real world it is me and this guy here learning about each other and studying each other around this fire, under this sky, sharing this breath of air and this earth we sit on, noticing that splash of sound from the nearby shore....

We stand and walk quietly to the water's edge, where a waterbuck glides out into the moonlit waters. Through my small binoculars we can see her dark outline as she watches us watch her. She sees more of us than we see of her, I'm sure.

Yeah, this is the real world.

We wander back to the fire and enjoy each other's quiet companionship for a short while longer, then I crawl into our tent, our zipped-together sleeping bags, and hold my wife, kiss her strawberry hair. The moon is a silver-white disk frozen in explosion on our tent roof.

✦ ✦ ✦ ✦

I admit, at this point, more than a week in Africa and several days into a wilderness safari, with lots of time to think and explore and have my mind expanded by novel experiences, I am feeling pretty good. Pretty confident. Pretty enlightened. I'm beginning to think that maybe I'm some kind of renaissance man, open-minded, forward thinking, fearless. Then the truth comes out.

I am pulled from the depth of sleep unwillingly, violently. I am shot awake by the staccato rhythm and monotone drone of a long monologue. I notice Sarah is awake, too. She grabs my hand and squeezes, but she doesn't move even though I can see the shine of her open eyes. Through the thin tent fabric I can see O.B.'s silhouette against the fire, raging tall and wild like a ceremonial pyre. Perhaps it is. O.B. makes flowing movements with his arms and body that I can't completely make out. And his voice, deep and serious, rides the huge flames, floating out words I don't know. I am afraid.

Then it ends. The fire is dowsed and he is gone. I hear him crawl into his tent and the million sounds of the swamp fill the void left by his voice. Except in my mind. It races with thoughts, fears run unleashed and out of control, ingrained images and ideas crash about like cornered cattle driven wild by a storm. I try to throw ropes of knowledge and understanding and experience around these escapees, these inbred critters of my society's racial and cultural myths and paranoias, but they run scared of the unknown, fearful of the different. *There's a savage out there conjuring up evil gods!* they scream. And for a moment, I believe them.

But I get them under control. I ride herd on these spectres and corral them into my subconscious, the only place I can put them. My heart stops racing and settles back into my chest, relaxing, like the camp watchdog calmed down and curling back up on his bed. I know that no matter how much life teaches me about the diversity of possibilities in humans and the world, this ghost herd, the Herefords of guilt, will escape to stampede and overgraze my mind again. I must accept that and deal with it. I know because before I closed the barn door to my subconscious, one straggler wandered off still free, bleating the question: "What's a good New England redneck like you doing with an upper-middle-class suburban Chicagoan like Sarah?" Then it was gone. It takes me an hour to fall back asleep.

◆ ◆ ◆ ◆

Dusky daylight filters a dull blue through our tent walls. I am awakened by... something. Sarah wakes, too, startled.

"What was that?" she whispers.

Then we hear it again: a groan, growing, like a bubble of sound, close and resonating, a deep and throaty *wwwhhuuUUUUHHHHuummpph...!* that trails off. Smaller, reinforcing groans follow, then the cycle repeats.

We lie in bed and stare wide-eyed at the roof of the tent.

An hour later, with water in the blackened pot on the coals, we ask O.B. what the hell that noise was.

"It is lion," he says, sculpting the coals around the bottom of the pan.

Like, what did I expect? This is the real world. O.B.'s words evaporate in the morning air as Sarah and I study each other's faces. Breakfast is silent. I sense O.B.'s casualness is a front; since I'm a guide, too, I know it's the guide's job to maintain a calm facade whether you think that next stretch is runnable or not.

Safari time again. The day is hot, and the African winter sun a great weight. We venture deep into the island that is scattered bands and stands of trees and endless meadows and open plains. Not far from camp we encounter a surreal sight: across a vast grassland a mixed herd of impala, zebra, tsesebe, and wildebeest flows along, turning and running in harmony, all caught in the same current, sharing the same mind. Quickly the herd turns to the left and circles behind us.

O.B. stares with us, but his thoughts run along practical lines. He nods, pointing with his head.

"Lion is there," he says.

We offer no reply to this tidbit of information. With his field guide books in hand, O.B. wanders on, veering slightly away from the direction he nodded in. Sarah and I follow closely, unsure of the way back to our tents. We trust him implicitly. We have no choice.

O.B. stops. We have walked a mile or so from where we watched the herds. He bends down and points out lion tracks, three of them in soft dirt. They are about the size of both my hands laid flat.

"This morning," O.B. observes.

We follow the tracks (in the opposite direction of the

lion's travel, thank you), and soon come upon a strewn group of large bones, many over three feet long, old and sun bleached. O.B. tells us they are the remains of a lion-killed giraffe. A lion, he says, can take any animal but a grown elephant. He seems to be as impressed with this fact as we are.

We continue along the trail of the lion, O.B. following the often-indiscernible tracks without hesitation. Twenty minutes of walking brings us to a grove of trees—and our camp. The lion tracks circle within fifteen feet of our tent. O.B. beams at us as Sarah and I absorb the distance, or lack thereof.

Nervous? You bet. Stunned, dazed, in awe. But after these past days with O.B., we aren't scared. We have grown to trust O.B., and that trust allows us to appreciate what is here, to see it for what it is—dangerous, real, and beautiful—with the sense that we are, as much as we can be as mere visitors, part of this place.

Trust is the tool of the trade of a guide. A broad knowledge of geography and flora and fauna and history are useful, but they are mere entertainment. What a guide really offers is a bridge to a place. The guide's comfort and relationship with the place allows visitors to follow in the wake of his passing, for a little while, at least. And when a guide achieves that level of success through which he can allow others to sample his understanding, his presence, the relationship he or she has nurtured with that place, then that guide earns a badge: an aura of effortless professionalism, of relaxed control, of peaceful alertness. An aura of trust.

O.B. makes us feel welcome here. What more can a guide do? I have dedicated a chunk of my life to guiding, to sharing, to creating in others some appreciation of the places I love. But now, for the first time, as I follow in O.B.'s wake through the African bush, I understand. I understand as I watch O.B. bury the ashes of his fire with almost ludicrous care, and as he pulls debris away from the base of a log to show us new-born mice, then painstakingly replacing the roof of wood chips and leaves. I understand as he laughs at our dazed realization that a damned lion walked by our tent last night.

O.B. has guided me to *place* as more than just the physical landscape; he has silently led me to a sense of place as the landscape that lives between my self and whatever is out there, to the landscape created by the relationship between the two. This is wilderness. This is where real wilderness guides work.

✦ ✦ ✦ ✦

That night, our last night in the delta, our last night in the swamps, our last night with O.B., we sit around a fire and sip Ricoffy. The sun lowers itself behind the horizon. O.B. smiles at the sunset, then waves.

"Bye, sun," he says.

2

KEN'S SHORTCUTS

*K*EN'S SHORTCUTS, MY WIFE CALLS THEM. AS IN, "OH, NO. IS THIS ANOTHER ONE OF KEN'S SHORTCUTS?"

These roads always seem longer, the ride rougher, the turns tighter than on our normal routes. They may not save time, but they always offer better scenery.

I'm not sure when I started taking them, but I think it was when I rode a purple stingray bicycle, long before I was issued a driver's license. I used to cut through a piece of woods and splash through the corner of a swamp to get to the Lake Boon General Store. My friends took the road, and usually beat me to the sodas and candy. They thought I was nuts. Now they own mountain bikes.

My father was probably an influence. Every now and then, on a Saturday, he would say, "Let's get the truck stuck." Simple as that. We'd pick up one of his friends, and the three of us would head out of town in my father's two-wheel-drive 1967 Ford F250. He'd plug in a Tom T. Hall 8-track tape, and we would try to get lost in the hills. We'd roam around aimlessly, stalking the worst road possible, until he and his friend would get out and push while I stretched my leg and burned the clutch, trying to get the truck rocking.

That's how I learned how to drive. I thought my dad was nuts. I still do, in fact. Now, I own a four-wheel drive, and rarely push.

I do often roam around, though. And roaming, of course, was also the true purpose of those Saturday meanderings with my father: the goal wasn't really to get stuck, it was to explore, to find new ways to new places. The fact that we often got stuck says more about how much he liked the exploring, not the pushing.

Today I may not look for the worst road possible, but I still seek forgotten routes that nobody cared enough about to improve. I look for narrow roads that circle around the smallest hills, that parallel the meanders of creeks and rivers, that have never had their personality smothered in pavement.

And if this route connects two places I need to go anyway, then it's a shortcut, even if there's a straight, flat, wide, smooth, "improved" road connecting those two places efficiently and directly, and at great taxpayer expense. I'll drive through old neighborhoods, up narrow valleys, over forgotten passes, and around mountainsides just to get a sense for the lay of the land, to see a new view, to relish the feel of wheel in hand and elbow hanging out the window.

Road builders today are too good at what they do. Cut-and-fill has turned topography into a backdrop, seen but not felt. Constant paving and road improving makes driving as effortless as watching T.V. It is so fast and easy to get to places that we forget what lies between them.

Driving will never be like walking, but when I drive I still want to sense the land I pass over. I want to feel like it took at least some time and effort to get somewhere. I want to feel more stimulated by what's around me than if I were at home watching the Discovery Channel.

So on Saturday, when Sarah and I were driving along the route we routinely take to a place we frequently go, I suddenly clicked on my blinker, and turned left.

Sarah glanced at me. "Where is this a short cut to?" she asked.

"To awareness," I answered.

3

DARK CANYON:
A WALK ON THE NEW FRONTIER

*A*T THE TRAILHEAD, WE SIGN THE FOREST SERVICE
REGISTER, ONLY THE NINTH AND TENTH PASSERS-BY
to log in during all of April. Above the register, the Forest
Service has left a helpful list: emergency phone numbers for
the nearest sheriff, fire department, and ambulance service.
These are all located in Monticello and Blanding, Utah,
nearly fifty dirt-road miles away. Where, of course, the nearest phones are.

We don't copy down the numbers. Instead, I close the register box, and Tom and I shoulder our packs and saunter out
into the huge, folded landscape before us.

My friend and I are headed into Dark Canyon, the wildest
of the wild heart of Utah's canyon country. On the map,
Dark Canyon lies like a god-sized image of a tree etched into
the Dark Canyon Plateau, south of Canyonlands National
Park. The thick trunk rises east out of the Colorado River
Canyon, and the many-branched top reaches 25 miles into
the high and remote Elk Ridge. Our walk will take us more
than twice that distance through the canyon's meanders,
from upper branch to lower trunk at the Colorado River.

Our route will also take us through the Dark Canyon Wilderness Area and Dark Canyon Primitive Area, the legal labelings of this remote place.

Remote, but not unknown. We're on the trail only an hour before we cross paths with three other backpackers. They're headed up onto Elk Ridge, where we just dropped from. They are ski patrolmen from Summit County, Colorado, they tell us, and are passing two post-ski season weeks walking the canyon country. We chat for a few minutes, as desert hikers do, about water, trails, routes, and the season's past powder days. After a few minutes the conversation lags; we've been lingering too long. None of us feels comfortable with this crowd out here.

"Did you bring your watches?" one patrolman asks, unexpectedly.

"Watches? We don't need no stinking watches," I reply.

They laugh. Although I'm not sure they caught the old-movie pun, they share the sentiment. We part, heading in opposite directions but toward the same spiritual destination, wishing each other a good trip, good luck, good water.

In the next hours, we drop from bench to bench, from ponderosa stands into piñons and junipers, and in and out of the dry wash of Trail Canyon, a name left over from when this was the cattlemen's route in and out of Dark Canyon. Above us, shimmering like the northern lights, is the white escarpment of Cedar Mesa sandstone—the same raw material from which are cut the stone archways of Natural Bridges National Monument, 20 miles south—a wavy curtain draped from the plateau country above.

By late morning we reach the floor of Dark Canyon, a wide intersection of side canyons with rolling sagebrush hills coursed by washes and gullies. We pick a gully and head downstream over dry sand and broken-stone pavement, until we reach the main-stem stream, and then—surprise!—a real, flowing, gurgling, sliding stream. Under a big cottonwood flapping lime-green springtime leaves, we stop, drop our packs, and suck up the shade like sponges.

Our rest is blessed and refreshing. We toast water bottles to our good fortune: trapped in the grasp of 1,500-foot ramparts with no escape for five days, until we climb out and walk the rim back to our truck.

"We are fools, *compadre*," I warn Tom. "We could be catching up on our work."

"You're the fool," he replies. "You're the one with a job."

Another toast. This celebration, coupled with water, shade, and camaraderie, pumps up our confidence, and we continue on.

The first miles of a backcountry trip are always like this: an intoxicating blend of energy and elation that numbs pain and the elements. Like all things, this, too, will pass. And soon it does. The desert heat, even in April, soon weighs on us like an anchor dragging in the sand, and the buzz is gone. A mile from our idyllic lunch spot the creek sinks into the ground, and the now-dry wash's cobble of rounded stones makes walking a chore.

We trudge, sweat, and ache. The straps of my pack carve canyons into my shoulders. The soles of my feet are raw nerve endings. After four miles I begin to ponder where, exactly, our little creek went, and when it might be back.

And once again, as on every other long pack trip, I confront The Question: *Why do I do this?*

✦ ✦ ✦ ✦

In the evening, we fill our water bottles from a tepid, silty and shallow little pool. A frog the size of my thumbnail eyes us warily as we drink his pond down an inch, then take a little more for dinner and tea. We apologize, and give thanks for the water. It is rank, yet every ounce is as precious as a hundred dollar bill. More so, out here.

We're camped at the junction of Dark and Woodenshoe canyons. Looking downstream past the wide junction I can see the canyon narrow, a tall, sheer corridor dark and endless in the evening light. From here on, we are committed to our adventure, with no backdoor exits for 25 miles. And even then, the side-canyon exit we are eyeing is uncertain. ("Yeah, I heard of someone walking out that way once, I think," the ranger said over the phone).

Around our little campfire, happiness returns riding a wave of adventurous spirit. But still, swells of dull concern swirl: How far to good water? Will my old boots hold out? Can we get out Lean-To Canyon? I lie down next to the red coals of our dying fire, the taillights of another day, and think of my wife, home alone. I think of myself, here without her.

Why do I do this?

The question keeps nagging, but I know the answer.

I do this to stay strong in body and soul. To keep my mind wild, creative, and confident. To learn through challenge. To discern real needs from luxuries. To flush urban culture from my system. To remind myself that there is, still, somewhere out there, wild country where I can rekindle my spirit.

A satisfying answer. But I, of course, am not the first to say these things. I bend open the ragged paperback I hold in my hands, and flip pages in the thin yellow firelight until I find these words:

> The wilderness masters the colonist. It finds him in European dress, industries, tools, modes of travel, and thought. It takes him from the railroad car and puts him in the birch canoe. It strips off the garments of civilization and arrays him in the hunting shirt and moccasin.

Historian Frederick Jackson Turner explained why I'm in Dark Canyon a hundred years before I threw my pack in my truck and headed here. His 1893 thesis, a copy of which I carry on this walk, outlined how European-stock Americans were molded by their newly appropriated land. Americans, Turner said in "The Significance of the Frontier in American History," need wilderness to be Americans. For us Euro-Americans, our personality and self-image is derived from having adapted to the wildness of this adopted continent, which was so different from the civilization-smothered one we left behind. The frontier, he said, kept these new Americans from being just so-many-more Europeans.

"American social development has been continually beginning over again on the frontier," he stressed. "This perennial rebirth, this fluidity of American life, this expansion westward with its new opportunities, its continuous touch with the simplicity of primitive society, furnish forces dominating American character."

Turner's late-19th century message was a mixed one, though. While he celebrated the role of wilderness and the significance of the frontier, he warned that the frontier was gone, a status confirmed by the 1890 census.

For myself, looking at America today, 100 years after Turner, it's hard to see much difference between the New World and the Old. We are all caught in the eddies of cities and autobahns and business and telecommuting. People in Denver, or even in relatively remote Durango, Colorado, where I live, watch the same T.V. shows as my friends in Boston do. Even in Sweden you can watch reruns of "Dallas." The BBC World News comes on the Southern Ute Tribe's radio station every night at 10 p.m. Without the frontier, I wonder, is there a distinct American critter any more? And I wonder, does it matter? Will the frontier mean anything to my great-grandchildren in 2093?

✦ ✦ ✦ ✦

The morning is long-pants and wool-shirt cool, but once the sun lifts itself over the south ridge, we're back into shorts and t-shirts. Out here, we roll with the changes.

And we roll on down the canyon. We repack and rehoist

and are soon threading our way down the rocky channel. I still wonder when we'll find a good, steady supply of water, but the full eight-pound gallon on my back—7.8 pounds liquid, 0.2 pounds silt—allays my concerns for now.

Sometime in the morning we pass from the Forest Service's Wilderness Area into the B.L.M.'s Primitive Area. We are not sure when or where we cross this boundary; the change is a legal one, a political delineation that the landscape ignores. To us, a couple of wayfaring foot soldiers, this is all wilderness without the capital "W," and no paper designation is needed.

Not needed, perhaps, but significant. That capital "W" means that people recognized the importance Turner placed on wilderness and the American spirit. So, some seventy years after Turner drafted his thesis, Congress passed the Wilderness Act of 1964. Today—a hundred years after Turner and on the 30th anniversary of the Wilderness Act—the frontier begins with a capital "W." It is static, its boundaries deliberate, chosen. It is the dark green on my map that stands out from the surrounding light green multiple-use area. It is Dark Canyon and its boulder-strewn wash, the spruce and fir hiding in shady seeps high on the canyon walls. It is the forty miles of dirt road we motored to get here. It is wondering where we'll find the next hole filled with dirty water.

It's not the American frontier of three hundred or even one hundred years ago, but this morning it feels pretty damn good.

And the good just got better. At my feet, water literally erupts from the sand, from under rocks, from the banks in a steady, fat, cold stream. In a matter of yards we go from a dry creek bed to a wet and noisy creek. We have stumbled on an artesian spring right where we needed it, a day-and-a-half into Dark Canyon. I say good-bye to the brown puddle water in my canteen and fill my belly with this gift from the ground.

Although we relish this treat, lounging and drinking and pouring water over our heads, the sun blazes on, and soon so do we. A mile downstream, though, we can no longer resist the water's call: We stop, strip, and dip. We dive into a deep plunge pool, a smooth bowl carved in the grey bedrock and fed by the steady pouring of nature's own faucet.

I float on my back, chilled but not quite cold, staring up at orange and grey cliffs framing an unblemished strip of perfect sky, like a blue filet mignon, a choice cut. I am joyous. Desert surrounds me, above and all around, yet I float in water. Miracles. Miracles are everywhere here, every day. Every minute walking in this canyon I peer into the raw, exposed bedrock and see...

...and see my own home. I see my wife and child and friends and community and little town and the marvelous miraculous landscape that holds it all.

I am surprised by these thoughts. Where did they come from? I re-examine: I was just floating here, butt-naked in this pool of desert water, thinking about miracles, when I realized...

I realized that miracles are everywhere, not just here in Dark Canyon, but in all life, even home-life, even in our urban landscapes. But I had forgotten. It's hard to remember miracles out there, back there, in the daily tedium of thought and logic, of hierarchy and economies, of smelly traffic and stinking watches. The glare of the urban world can blind us from the magic of kin and community, of life and land.

That urban world is today's equivalent of the Europe my ancestors set sail from. So I, too, come to the frontier, and in this plunge pool I wash off my dry, pasty-white European skin and emerge American once again. Again I see life and land anew: *They are all we need; they are all there is.* This is what the pioneer relearned.

❖ ❖ ❖ ❖

For the next couple of days, the creek is a constant companion. Sometimes it is wide and shallow and spills over ledges; sometimes it rides a deep, smooth gutter carved in sandstone. It frees us from the burden of carrying water, and offers relief from the heat. Miracles. Several times a day we pass pond-sized pools lined with cattails and sedges, and each night we are serenaded by frogs, burping and chirping, carrying on. The miles toughen my feet and strengthen my shoulders. The wilderness is chiseling away at my body, mind, and spirit.

> From the conditions of frontier life came intellectual traits of profound importance...that coarseness and strength combined with acuteness and inquisitiveness; that practical, inventive turn of mind, quick to find expedients; that masterful grasp of material things...that restless, nervous energy; that dominant individualism...which comes with freedom—these are the traits of the frontier.

We Euro-Americans may not always take pride in the way some of our ancestors carried out their pioneering, but we nevertheless cling to and hold sacred the spirit that pioneering passed on to us. Even if today we only pioneer our way through daily traffic snarls to our office and back, we want to live our lives with the pioneer's spirit.

The afternoon is hot when we reach a stretch of canyon that is notched and dotted with a string of deep pools. Tom changes from his boots to sandals to try walking the slickrock creekbed; I, the cautious one, keep my boots on and choose instead to head up and seek a path along a bench above the creek, threading my way along the terraced slope.

We head down canyon on our separate paths for about half a mile when I notice that, below, Tom's luck has run out. He stands, hands on hips, staring into a pool too deep to wade. He adjusts his plan. He scales the boulder-strewn bank opposite mine and picks his way downstream. Soon, though, his luck runs out again, along with his rock shelf.

On my side of the creek (the passable side), I stop and watch, setting my pack down and nibbling dried apricots. The plot thickens. Tom ropes his pack down to the next shelf, and after much deliberation, makes a risky jump to his pack. Success, but it is short-lived: this lower level is also a dead end, and the lip he jumped from is too high to get back up. An interesting predicament, I think as I dig out a Zip-Loc full of gorp. In canyoneer's lingo, my friend is "rimrocked."

Tom walks back and forth looking serious. Then he finds it: an ancient and weathered juniper trunk is wedged part way up the wall from the shelf below to his own dead-end bench. He ropes his pack again, then lowers himself down on the log. His leg, foot, outstretched toes don't quite reach, but then he releases his handhold, drops, and lands on the top of the trunk, which he monkeys himself down. I yell a cheer; he slides his pack on and scales his way down into the creekbed, now below the deep pool.

+ + + +

The air is thick and hot when we arrive in the lower end of Dark Canyon, four miles from its junction with the Colorado River. We're at 4,000 feet now, 4,000 feet lower than where we started and at least a month closer to summer in climate. Here, the cliffrose and canyon holly are in bloom, and the ground is speckled with wild rose and sego lilies. We camp at the mouth of Lean-To Canyon, which we will try to use as our exit out of Dark Canyon and onto the Dark Canyon Plateau. We both know it may only offer us unscalable pour-offs.

No matter for now. We rest a couple of days under sheer red cliffs, great palettes for desert varnish tapestries, and backdrops for deep-green cottonwoods. We pass one day walking packless down to the Colorado River, and there ritualistically submerse ourselves in the muddy blood of the West's heartline. The rest of the time we relax, read, talk, swim, smoke big cigars by the fire and look up the narrow depths of Lean-To, wondering how it will treat us.

+ + + +

Lean-To treats us harshly. Our first obstacle, which we confront in the grey of early dawn, is a 400-foot overhanging pour-off. We gain the top by warily scaling the east wall of the canyon, climbing from shelf to narrow shelf, crossing steep chutes that end in broken talus on the canyon floor. If every day began with such a task, I think as I at last peer over the overhanging lip, I would never have to drink coffee again.

The next several miles present no more giant hurdles, but instead batter us with what seems like a freak of nature, a surreal landscape, another in the desert's endless stream of surprises: We must hack our way through thick underbrush, marsh, swamp, and hidden mossy deadfalls under a green canopy of foliage. I think of "Land of the Lost," of Indiana Jones, and of Central American jungles. I think of everything but Utah's canyonlands, except when I look up through a hole in the thick, overhanging cottonwood ceiling and see the familiar redrock-against-blue-sky backdrop still framing our world, no matter how weird it has turned. Damn miracles, again. But with my bruised and cut shins, muddy feet, aching back, sweat-soaked brow, this miraculous place will be one best remembered rather than experienced.

After four grueling hours, we break out into familiar, traditional canyon country again, a sandy channel choked with boulders and supporting a thin trickle of water. We limp up the canyon a few more miles, our bodies beaten from the morning's bushwhacking. With our last strength, we push ourselves up, out of the canyon and onto the rim, where we see for the first time in a week the grand landscape around us: the orderly row of the Henry Mountains, the purple gash of the Colorado River canyon, the innumerable red hummocks and buttes of The Maze. All the world is dazzling and sunlit except for one dark cloud in the distance trailing a drift net of rain.

We at last drop our packs under a rock overhang as a fork of lightning kisses a canyon rim, and we fall asleep as thunder rolls by, echoing across the huge space like a bowling ball in a kettle drum. I dream of my wife.

✦ ✦ ✦ ✦

The last leg of our journey takes us some twenty miles on a two-track road across the Dark Canyon Plateau, a land-bound peninsula fingering out from Elk Ridge to the Colorado River gorge. For a day and a half we walk, talk, think, talk, and walk. I am sore and tired, yet eager and energetic. The rhythmic creak of my pack is a drill sergeant. *HUT-two-three-four-HUT-two-three-four....* I am barnsour as an old pack horse.

Every aching step (the last miles for these old boots) reminds me of Lean-To's backbreaking ascent. It was fun, but not many people would enjoy such hacking and scaling. I'm not sure even I want to wander up that way again. Still, even if I never peer up Lean-To's ominous slot again, even if no one ever discovers its misplaced marshes again, it's important that forbidding places like that are out there, just in case. It's important not just for those of us fool enough to enter, but for those who merely want to gaze from a distance, or who are happy to imagine over the contours of a map, or who will only hear the tales over a bottle of beer in a Denver bar.

Why is it important? Because just the existence of these wild places seasons the spirit of all Americans—not just the European descendants—creating a distinct flavor that is uniquely American. As my friend Turner put it:

> The most important effect of the frontier has been in the promotion of democracy here and in Europe.... As has been indicated, the frontier is productive of individualism.... It produces antipathy to control, and particularly to any direct control.... The frontier conditions prevalent in the colonies are important factors in the explanation of the American Revolution.... The frontier individualism has from the beginning promoted democracy.

Frontier and wilderness breed revolutionaries, libertarians, anarchists, and democrats in the truest sense of the word. Seasoned with wilderness, the American spirit is hot and spicy. If we want to keep that spirit from turning into over-civilized and over-domesticated white bread, we must keep some—a lot of—America wild. We must have wilderness, designated Wilderness, the new frontier.

But I fear. Too many Americans today don't know, or don't care, about the wild side. Even though outdoor recreation grows in popularity, wild areas continue to be devoured. Golf courses are justified as "open space." National Parks continue to be managed as amusement parks. Endangered species are either nuisances or a fashion statement on t-shirts. A paved road is proposed to the rim of Dark Canyon.

✦ ✦ ✦ ✦

On our last night out, the night before we reach the truck that will carry us back over dirt roads, to pavement, to towns, to work and watches (and, yes, back to my wife and home), Tom and I finish our last cigars and talk over a sweet juniper fire. We discuss our children-to-be, and their children. We wonder aloud if in a hundred years, on the 200th anniversary of Turner's thesis, the 130th anniversary of the Wilderness Act of 1964, if people will be able to walk away from the demands of their world and revive their wild senses, their wild minds and spirits, become Americans again, in some kind of frontier. Will those who want to tackle a Lean-To Canyon be able to find one without a road or sidewalk or fenced overlook or mine or nuclear waste repository or country inn? Will those who just want to walk to the rim of Dark Canyon to peer over, look in and take a deep, fulfilling breath just to be assured it's *there*, be able to?

Somebody must make it so, we agree. Somebody must make sure.

✦ ✦ ✦ ✦

As we walk the last miles to the truck under a swirling

cloud-filled sky, Tom, a Latin-American historian, tells me the story of two Mexican defenders from the frontier, from the wilderness, Pancho Villa and Emiliano Zapata.

"All they wanted was to live off the land," Tom lectures, pumping his fist into his palm, "in the same communities their grandparents had lived in. Politics and industry had other plans. But they wouldn't move. They fought back."

Those two knew the truth about life and land, kin and community: they are all that matter. I pull out my dusty copy of Turner and read to Tom:

> In this connection may be mentioned the importance
> of the frontier, from that day to this, as a military train-
> ing school, keeping alive the power of resistance to
> aggression, and developing stalwart and rugged quali-
> ties of the frontiersman.

Yes. We need more than savvy environmental lawyers and economically strong outdoor enthusiasts. We need unbending defenders not just *of,* but *from* our chosen new frontier. Only they can remind us that the frontier was, is, and must remain more than a playground or a zoo or a tourist attraction, that wilderness justifies itself spiritually, not economically. Only these true Americans can show us that we need the protection of wilderness as much as wilderness needs our protection.

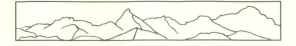

4

GOOD NEWS

*T*HERE'S BEEN SOME GOOD NEWS LATELY: THERE IS STRONG EVIDENCE THAT GRIZZLIES STILL LIVE IN THE South San Juan Mountains; the San Juan River is still a place where Colorado squawfish can live and breed; and, most recently, a federal study concluded that Colorado can support more wild wolves than is proposed for reintroduction in the entire Idaho, Montana, and Yellowstone areas combined.

In and of themselves, these are just a series of separate events, and that is how they have been reported in the papers and on T.V. Just so much more news. What makes these things good news is what they mean: They give us proof that we live in a place still wild enough for top-of-the-food-chain predators.

Together these events comprise an unreported news story of *hope*. Good news, compadres.

How do wolves and grizzly bears and squawfish add up to hope?

The interior West, between the Sierra Nevada and the Front Range, is one of the few places in the entire industrialized Western world where this news could happen. Only here is the ecology still alive enough, the wildland base still large enough, the human population still sparse enough to support the big predators. Europe has some wild dogs, a handful of harassed wolves may still cling to survival in Scandinavia, the East and West coasts harbor small populations of wildcats and bear. Northern Canada and Alaska are still refuges for wolves, grizzlies, and salmon, but these places are outlands, not as plugged into the main circuitry of industrial society as the interior American West.

We are it, folks.

The presence (or, at least, the possible presence) of these large predators indicates that the landscape's ecology below them is intact and healthy enough to support these top-end, specialized critters. Just like in the human body, if the vital organs suffer major trauma, the body progressively shuts off the major extremities—the limbs lose blood flow and body heat and usefulness. The body may be alive, but it is incapacitated. This is wild land without the full use of its major limbs, the predators. It is ecology on last-ditch survival mode, not fully functioning. It is dying.

Grizzlies and wolves and squawfish (the top predator in the Colorado River system) mean health, and health means longevity, if we take care of it. What these recent news events together add up to is the news that we still have a wild, thriving, living landscape, a place we can ourselves enjoy and also pass on to future generations.

I know that this can be easily lost. I grew up in New England, a heavily developed place that was wild as recently as two or three generations ago. I heard only tales of wild animals, of big wilderness. I hunted and fished and walked all over New England, but being on that landscape felt like being in the company of an old man telling deathbed stories of a rich life I will never experience.

I felt ripped off, like I was born too late.

Wolves, grizzlies, squawfish—these are the vital signs of a living wild land. The signs are weak, but at least the American West has a pulse, which is more than most places can say anymore. There is hope; now we need to guard it.

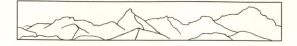

5

SOME THOUGHTS ON GROWTH

*O*KAY, IT'S TIME I COME OUT OF THE CLOSET: I AM ANTI-
GROWTH. I WANT TO KEEP IT LIKE IT IS. KEEP IT LIKE
it was.

I hear the cries already. *Burn the witch!* This is such a vile
stance to take that even the most hard-core environmental
groups and activists avoid it, tacking onto their proposals the
disclaimer, "We're not against growth, but...."

I hear the first retorts: Growth is inevitable; it happens
whether we want it to or not. Okay, let me clarify my position:
I am against actively promoting growth in our community.

I speak this view from experience. I've seen it happen.

So now I must admit Cardinal Sin #2: I moved here.

I moved here from New England, where I witnessed in the
late 1970s the "Massachusetts Miracle," an explosion of growth
from a promotion of high-tech industries that would make the
economic boosters in this area salivate all over their wing-tips.
High-paying jobs were there for the picking, and rural land was
snatched up at inflated prices like life jackets on the Titanic.
Politicians, business leaders and the media sang and danced.
Growth was a miracle.

Meanwhile, I listened to my parents and their friends gripe

and groan about the new-comers and the "aginners," people protesting the great changes rural New England was undergoing. My folks got good jobs from the growth. A few years later, my parents even made a little money off the boom, selling their house at a ridiculous price. They sold it so they could move away, mourning as they headed south the loss of the community and countryside they were born and raised in.

And now, here in the interior West I hear the same lines, the same singing, the same griping and groaning.

An economy based on growth is good only when measured by narrow standards: short-term monetary gain. But life and community are more than that. They are the health of the environment, room to breathe, affordability, stability, knowing your neighbors. Quality of life is the knowledge that you can pass those things on to future generations.

Promoting growth in a community works against those things. It means short term gains for a few, and long-term losses for everyone else. Remember: Development is forever.

I hear the next retort: No growth means death. Not so. All living things grow in infancy, mature and learn, then reach a healthy, sustainable size. That size is nurtured by caring and maintenance and through a respect for natural limits. Exceeding that size leads to unhealthy obesity or deathly elephantiasis in individuals, and disease and sickness in populations.

Unchecked growth is not natural, not healthy, and to seek it deliberately is insane.

Children in the throes of adolescence put their self-interest and immediate gratification above all else in their thinking and actions. They place the present over the future, fun over joy, immediate reaction over thoughtful wisdom. Although this is an important and vital stage of human development, it is only a stage; we must mature to survive. And no sensible parents would let an adolescent run their family's businesses and affairs.

So why do we let adolescent economics—a growth-fueled economy—run our lives? Why do we let business and political

leaders with the values of children rule our communities?
Growth is inevitable? Question it.
Growth is necessary? Don't believe it.

6

WIDENING TRAILS

*R*OCKY MOUNTAIN TOWNS TRADITIONALLY PUMP UP
THEIR ECONOMIES IN ONE OF TWO WAY:S: EITHER BY
digging holes or by widening trails. Here in Colorado's Fraser
River Valley, the latter has been the favored method. Without
much to mine except granite and glacial till, trails have been
the key to economic development.

The valley's main trail to money, and the recipient of most
widening, is the one I drive today, Berthoud Pass. This
twisted section of U.S. 40 is etched into the Continental
Divide, requiring 75 miles of pavement to wind the 40-mile
distance between Denver and Winter Park, the first town
encountered west of the Divide. In the past one hundred
years, this route has been transformed from a faint Ute foot
path, to a horse-pack trail, to a heart-pumping stage ride, to
a seasonal auto road, to a well-maintained paved highway
complete with passing lanes. Even with this most recent
improvement, the pass still crosses the Rocky Mountains,
and so is subject to snowfall, avalanches, and rockfalls. Even
in fine weather the drive is a challenge with its tight switchbacks and steep grades, demanding a little nerve and a lot of
shifting.

This is especially true in my little four-speed pickup truck, giving all she can of her two-liter soul. No hurry, though; it's a blessedly inspiring, early spring mountain day. Snow-white clouds yield to a lake-blue sky, and last night's snow turns to rivulets on the roadside. From Denver, at 5,280 feet, to the top of the pass, at 11,300 feet, my Ford and I climb more than a mile, more elevation gain than driving from Boston to Denver.

Despite the automotive challenge, Berthoud Pass is the main vein to the Fraser Valley's economic heart. This road connects a land of tourist bliss with those offshoots of the urban East that have taken root at the foot of the Front Range: Denver, Aurora, Boulder, Colorado Springs, Fort Collins, and their sprawling environs. That's where the people, and therefore the dollars, lie. To get those people to the valley so they can spend those dollars skiing, snowmobiling, fishing, mountain biking, hiking, and horseback riding, Berthoud Pass has been massaged, smoothed, paved, and graded. Improved roads for improved business. Widening the trail.

My truck finally lumbers to the top, between Mount Nystrom and Colorado Mines Peak, and there I find a threshold. Behind me roll the Foothills, grey and green with granite and pine; ahead, the view explodes in contrast. In that direction, the Divide is a rampart of steep, stern, humbling peaks stretching some 50 miles from the pass to Longs Peak, in Rocky Mountain National Park. Off this grand wall, folds in the valley fall like a bedspread to cool-dark valley floors. These folds, all small drainages, eventually merge their collective waters into the Fraser River, whose wide valley is a basin held aloft, tilted northeast toward the Colorado River, 25 miles away. In the distance, a jumble of peaks, hills, and open chaparral draw westward.

They draw me westward, anyway. I'm headed home.

Technically, I suppose, the Fraser Valley is not my home anymore. After five years eddied out from the mainstream in the three tiny towns on the valley floor, I moved back into the current of society, "over the hill" as Fraser Valleyers call the pass, to Boulder and school. But every few weeks I gas up and cross the pass. I have a need to cross that threshold, stand on the pass, and drop down into the world beyond. There's a spirit pooled up in this valley, flowing among the dwellers here, and this spirit suits me. I am an addict who needs a hit of this fantastic landscape and the communities built upon it, so my truck and I hum down the north side of the pass, joyous.

Joyous, but nervous. As I roll past the Winter Park/Mary Jane Ski Area and the two miles further to the main street of Winter Park—all 700 yards of it—my mind frets and my stomach knots. I keep thinking about the tunnel.

The Berthoud Tunnel. It seems widening the trail is no longer enough for the visionaries of economy, the land developers, the ski area planners. Now they want to dig a hole, too. The plan is for a four-mile, flat, horseshoe-shaped bore through the Divide replacing the 16 miles of Berthoud Pass's meandering roadway. The drive would take 20 minutes less, the pavement would be eternally dry, and steering and shifting would be effortless. Thanks to this feat of engineering, student drivers and Kansans never would again lose sleep anticipating the pass.

To the promoters, the tunnel would be a big straw. This easy, safe, and quick passage would siphon drivers off the Front Range cities and beyond, depositing them and their fat wallets into the valley's ski areas and towns. Like an incantation, The Magic Words roll from the developers' tongues: The Economy Will Grow.

That's the theory. But the tunnel promises to change more than the drive. With the increased auto traffic, the land speculation, and the invasion of businesses and builders, the towns of Winter Park, Fraser, and Tabernash, and the predominantly forest and meadow landscape of the valley floor, will pass forever into a new era. From the hole in the mountain will flow people and money, and out the hole will drain the rural, small mountain-town landscape and its spirit.

The story is a familiar one throughout the New West, where the old West's traditions of land-based individuality and independence (real and mythical, for the West has always relied on government largesse) are being displaced by a chamber of commerce mentality based on resorts, tourism, and service economies that strip mine culture, social structure, open space and living room. In places like Taos, Jackson Hole, Central City, Sedona, Santa Fe, Moab, and Telluride, trails are widening to greet this New West, but they're not widening gradually anymore. Today, it's dynamiting the dam rather than slowly opening the penstocks. If it's not the Berthoud Tunnel, it's a new airport, a new interstate, a new ski area. No one is happy just to be small, close and friendly—but second best. "The W.P.R.A.'s (Winter Park Recreation Association) primary management objective is to become Colorado's leading provider of quality mountain recreation," boasts the ski resort's expansion proposal. But don't all ski area's want this? When does it end? And what are the costs?

The costs are legion. The financial costs for the Berthoud Tunnel are still estimates, and those estimates have the usual problems found in big developments: contradictory figures, inflated population and use projections, risky financing proposals, dubious claims and statistics. But these anticipated costs and their problems at least get discussed. The media covers these costs and controversies because they're tidy and verifiable, real numbers spewed by authorities and experts, the developers, economists, lawyers, politicians, and organized opposition groups. It's the costs that don't get voiced in the media that bother me. I mean the cultural and social and spiritual costs that are non-verifiable, non-quantifiable, and less-well articulated by the experts in community, lifestyle, and spirit: the seasonal laborers, the family raisers, the woods-wanderers, the front-porch guarders, the simple-lifers, the root-sinkers. These are people less likely to speak up at meetings than they are to help their neighbors, less likely to afford the money for an ad campaign than they are to afford the time to walk up some small creek, who can't rattle off statistics but can remember meaningful experiences of the place in which they have chosen to live.

I don't doubt the tunnel *can* be built; I question whether it *should* be built. In that decision the perspectives of these other experts are just as valid—probably more so—as the reams of economic logistics and feasibility studies so revered in the media and public debate.

Today I want to talk to some of those experts. So in Winter Park I turn right across the street from the 7-Eleven and between a Conoco and a ski shop, and pull into a small group of trailers decorated with a big wooden cactus and bird silhouettes. This is the Winter Park Youth Hostel, cheap lodging for low-budget travelers. Many of the guests who stay here are European or Australian, visiting or ski-bumming America's mountains. This is the first place I set up shop when I ventured westward some years ago.

Inside the office trailer I find Polly and Bill Cullen, managers and owners of this 14-year-old establishment. In that almost a decade and a half, this simple business hasn't made these folks rich, but it has kept them fed and given them time to walk and fish and travel, and offered them a way to meet other kindred spirits. They are also the most vehement of the tunnel opponents. It wasn't an easy or popular position to take.

"We got pretty well crucified," says Polly, pouring tea for the three of us.

When they first spoke up, resisting the followers lining up to hail the tunnel as the Fraser Valley's economic savior, Polly says they were "patted on the head" for speaking their minds. But once the tunnel's proponents realized this couple was articulate, investigating the project's details and uniting other opponents into a cohesive group, they were ostracized in the community. Despite the ridicule, the Cullens persisted in their resistance because, they say, the developers "were threatening our lifestyle." Their quality of life was under attack. And what's that quality of life? Bill ponders a minute, then says that he leaves his keys in his truck. They often leave the door of the trailer open while they're gone, he adds.

And what would they receive in trade for that quality of life? One promise the tunnel pushers make is that the tunnel will make the Fraser Valley another Vail. This makes Bill wince.

"If I want Vail I can move over to Vail," he spits. The fact that this valley is not Vail or Aspen, with their dense developments littering those valley floors—the expensive ski shops, jet-set glamour and celebrity spotlights—is this place's ace-in-the-hole, the Cullens argue. If the Fraser Valley's undeveloped character was preserved until it were a rarity, it would raise the valley's value while maintaining its unique and quality atmosphere. "It's like putting away a mint coin when it's minted," Bill says.

Polly paces excitedly and Bill calmly sips his tea while they bombard me with statistics and economic figures they have accumulated. These call into question official estimates of traffic use—essential to showing a need for and income from the tunnel—waved by the proponents and regurgitated by the press. A high volume of traffic is essential to the tunnel financing plan in which drivers will pay a $3 toll for the tunnel even though the pass road will stay open and free since a federal highway cannot charge a toll. Bill's hand-drawn bar graphs in yellow, blue, and red show that in the last two years traffic fell short of estimates by between 15 and 60 percent at various times of the year.

Bill continues a logistical barrage for a few minutes, but then Polly interrupts him, all flustered. She wants *emotion,* she says. But, she moans, she has learned to not be emotional when talking about the tunnel because officials won't take that seriously. Only intellectual reasoning is accepted in the debate, she says.

"We can't just say we don't want the fucker," she says, angry. She tells me the story of a mutual friend of ours, an articulate and sincere guy but heatedly emotional about this valley he calls home and what the tunnel would do to it. His honest passion finally "became a detriment to us," Polly explains, shaking her head and frowning.

They may despise the tunnel, but both these tunnel-hinderers assure me they are not anti-growth. It's just that growth shouldn't be explosive, like the tunnel would ignite, they say.

"We need to offer more in the valley," Bill explains. "We watch cars whiz by all the time."

I leave the Cullens, climb into my truck, and head back onto the Winter Park strip. Bill's thought reverberates in my head: *offer more in the valley.* I look across the street and see the town's only real effort to get people out of their machines and on their feet, a tasteful wood-sided, three-story walking mall filled with quality shops and a night club. The rest of the town glares in contrast: the 7-Eleven (the showpiece of main street) where gangs of kids hang out (not in gang colors but in ski jackets and Lycra), gas stations, a few restaurants, a Pizza Hut, a grocery store, and a movie theater that has shown the same movie for three weeks now, all laid out in a widely spaced line. There are no bowling alleys, no bars for 18 year olds, no recreation center, although a group tried to get the town to build one a few years ago. Winter Park bills itself as a family resort, but I don't think the novelty of the longest road tunnel in the United States will lure these young skiers back. New but empty sidewalks line both sides of the busy street, an unaccepted invitation.

Why accept? Why walk? Winter Park is not Aspen or Telluride or Breckenridge, with their pretty settlements of Victorian mining homes and dense business districts. Those towns are a pleasure to stroll through and offer alluring and unique characters that make people want to visit, explore, stay a while. Winter Park's main boulevard is a gasoline alley, like those quick-stop interstate exits. It reflects the character of the resource mined here, traffic off U.S. 40: It's a pit stop rather than a vacation spot.

Does the Fraser Valley really have to stoop so low as to beg for auto traffic? I wonder as I roll through town in my auto. Does it really need to turn into Far-West Colfax, an extension of the dirty pavement-and-cinderblock business strip that is U.S. 40 through Denver? It's not too late to change that course, with time and money and thought and some affection for this place. The town is still small, more trees than signs, not yet overrun with gas stations, McDonald's, Qwik-Marts, Taco Bells, and Dunkin' Donuts. But this does look like the prime habitat for those alien species if something healthy isn't planted soon.

Leaving Winter Park, I drive northwest, toward Fraser. In this space between towns I am reminded why I moved here and stayed a while, and why I keep coming back. The Divide dominates the eastern horizon like a great looming and petrified wave. Looking north, out the right side of my windshield, I see the sheer-faced and serrated Indian Peaks, an escarpment backdropping the rumpled foothills that roll down to the smooth valley floor. Sweeping by my left window is a snow-covered meadow, the size of ten-or-so football fields side by side. A half-dozen shaggy bay horses paw through the snow under the shine of the western wall of the valley, the sun-lit Vasquez Range. Driving to work I have seen coyotes in this meadow, often two or three, romping and bouncing in the greening grass or heavy snow. Sometimes they would be just sitting together, facing east, pondering... something. The sun crawling over the Divide? The stony couloirs of the Indian Peaks? The traffic between Winter Park and Fraser? The ends of their noses?

This wildness so close to the ski areas, the condos, and the stores is a major lure. One real estate company's brochure proclaims: "Colorado's Winter Park: It's a Natural." Inside, the sales pitch continues, "Winter Park is Colorado's natural vacation land. A land of unspoiled beauty, abundant wildlife and friendly people." Realtors, though, make their money selling land. Maintaining the valley as a "natural vacationland" is not the path to profit. Another real estate advertisement in the local paper cajoles: "If you missed out on Vail and Aspen, don't miss out on Winter Park!" And away we go.

Not all realtors have jumped on the tunnel bandwagon, though. Bill Stewart, an area realtor for 16 years who has lived in the valley since 1970, has jumped in front of the wagon. He says the tunnel is an economic bum deal for the town. His opposition eventually led him to challenge the head of the Berthoud Tunnel Building Authority, another real estate developer, for his seat on the Winter Park Town Council. Stewart whipped the tunnel proponent 73 to 34.

Still, Stewart's active stance against the tunnel wasn't all that easy. He was, he says, "hounded on the telephone" by other realtors who saw him blocking opportunity. When the ski area, a key backer of the tunnel, purchased property for employee housing, Stewart got none of the business; the two realtors who did were the two most vocal tunnel proponents and most angry at him, he says. Even his wife "thought I was a dummy" for being against the tunnel at first, he laughs. But that has changed. "I even heard her on the phone the other day explaining why she was against the tunnel," he says.

Getting hounded on the phone and having your wife disagree with you is not the worst of what can happen if you swim against the popular current of the powerful in a town of 700. A local ski shop owner threatened to fire any employee opposed to the tunnel. An advertising company run by a vocal tunnel opponent lost half its income when its biggest contractor pulled its business. In a much-publicized incident, a ski patrolman and his wife were evicted from their apartment of seven years for their anti-tunnel views. The landlord, the father of the tunnel authority's director, explained in a letter to the local paper: "Since they have expressed feelings that are adverse to the things our family has worked for for many years, we felt it was time to make changes.... Who we rent to is important to our peace of mind."

I continue on past the big meadow, past the new Safeway, and into Fraser. Although this town of about 2,000 has no say in whether the tunnel gets built or not—Winter Park alone will be financing the project even though it has been predicted to affect communities as far away as Kremmling, Steamboat Springs, and Craig, on the Utah border—the issue is a big one here. The economic boom expected to flow out of the tunnel and all over the valley would change this town immediately and forever. (One angry and inebriated critic at a Fraser drinking establishment labeled the tunnel "the Berthoud Butthole" for its obvious physical similarity, but more for the quality of development he felt would flow out of the hole and into the valley.) The tunnel has also aroused animosity in this town. According to a Fraser town council member, the pro-tunnel mayor suggested all council members opposed to the tunnel resign. They didn't.

Driving through Fraser, it's obvious the town suffers from an identity crisis. The dirt side streets are lined with an odd mix of old log cabins, dusty groups of tin trailers, beautifully restored historic homes, and cheap pre-fab box houses. On the edge of town are recently constructed condominiums, six powder-blue two-story rectangles lined up like barracks.

Regardless of the housing potpourri, this town still has the closest thing to an authentic, historic atmosphere in the Fraser Valley. Fraser was once a rough logging town where worked German POWs and lived loggers who beat up the wimpy, faddish skiers. A number of sturdy and historic buildings stand around town, still lived in or housing businesses. The dirt streets are more atmosphere than annoyance. And a local bar as recently as a couple years ago still had a mountain-style rough-and-tumble atmosphere at its Wednesday "Quarter Draw Night," locally known as "Wednesday Night at the Fights." Some people, the heartier tourists and the simpler locals, like myself, found this more charming than dangerous, and more authentic than some tourist "saloon." And the beer was cheap—try to find that in Aspen or Vail.

Today there is evidence of the town's efforts to capitalize on its history, with old-fashioned wrought-iron streetlights and a town-center sculpture of a bucking bronco (Fraser is on the state rodeo circuit). But there are also portents of a not-so-tasteful future. On the edge of town sits a new shopping mall, an Arby's, a KFC (which, to its credit, did for a while maintain some mountain-town ambiance—for years it was encircled by a rutted, muddy route to its pickup window, which locals proudly proclaimed as the country's only "Four-Wheel-Drive-Thru"), and nearly a dozen land-for-sale signs.

Foreboding signs. But for now Fraser is still small and plodding hesitantly into the uncertain future. It still offers unique and valuable non-quantifiables, that ethereal "quality of life."

"It's an ideal environment for a four year old," explains 21-year Fraser resident Lynda Muhlbauer. We sit in a local restaurant sipping coffee. With a pretty, mountain-weathered face and a 95-pound frame, the emotional force she delivers when she talks about her home is surprising.

And the effects of the tunnel?

"Totally negative," she declares. "I don't worry that anyone will kidnap my child here," she says. She looks around the restaurant. "I know everyone in this room."

Fraser needs slow, steady growth, she tells me, echoing the Cullens' sentiments. The challenge to the town, she says, is "can we build without losing our lifestyles?"

Or without losing their livelihoods. The tunnel has inspired a type of gold fever, she says, and the heat has made the business people irrational.

"They don't see that we won't have to buy from their ski shops anymore," she explains. "I would love to blast to Denver in a half-hour and pay eighty-nine cents for a dozen eggs instead of a dollar forty nine, but do that and soon they'll say we don't need a Safeway up here, and we'll lose those jobs."

We leave the restaurant, head back into the fine spring air, say good-bye. I need a walk, need to sort out these thoughts, need to remember why I come up here, why I came up here that first time several years ago. I leave my truck in the restaurant parking lot and head to the Denver and Rio Grande West Rail Road tracks that run like a spine down the middle of the valley.

In the years I lived here, I won and lost several loves in this land between the Front and Vasquez ranges, and I owe a lot of broken-heart mending to the miles I walked on these railroad tracks. I don't miss the broken hearts, but there are a lot of things I do miss in this place. I miss the arctic winters when I would hitchhike to work in temperatures under 40 below zero and while snow fell from the clouds my breath created. I miss seeing the rocky spires of the Indian Peaks under the first September snow, and the morning sun setting fire to the top-hat top of Byers Peak. I miss outrunning thunder-snowstorms in May, getting 4X4's stuck in snow drifts in June, and in July taking softballs in the chops off the corrugated softball fields.

I even miss the tourists who invaded the valley. I passed several summers floating customers down the Colorado River west of here, where it cuts through the Gore Range. One time as I drove our people back from the river, a family sat behind me. As we rumbled along the highway the father leaned toward me while his wife and daughter slept next to him, as most people do after an active day on the water under the sun.

"That river was the most beautiful place I have ever seen," he said. "Thank you."

I even miss those other visitors to the valley, the ones who left their brains at the office (and why shouldn't they?). I mean the ones who, in my bus-driving days, were still sitting silently in the back of the bus when I pulled into the ski area after a 50-minute run through Winter Park and Fraser.

"Um ... where are you going?" I asked timidly.

"To our condo."

"Which condo is that?"

"The red condo."

"Which red condo?"

"The one you picked us up at this morning."

I hope they made it home okay.

I miss seeing each ski season's crop of young newcomers come to ski bum and break their programmed, pre-ordained, mainstream school-to-work-to-marriage-to-family track. I miss seeing the beauty and simplicity of life in the Fraser Valley rock their ingrained world-views, like mine were. Suddenly land and community and really living take on a vital importance; they learn that being irresponsible is not leaving home and ski bumming, it is doing things you don't really want for reasons you don't really believe. Many stay here. Most who leave, leave changed.

So what's ahead for this valley that keeps a firm grip on my spirit over many miles and mountains? I sit on the railroad cinders and stare out toward the meadows between Winter Park and Fraser, toward the highway dividing the meadows, toward the Continental Divide just barely separating the urban Front Range from the rural Fraser Valley, and I try to imagine this view in twenty years if the tunnel is built. I see a beltway conducting an endless stream of traffic around an endless business district of chain gas stations, chain convenience stores, chain fast-food restaurants, chain ski shops and factory outlets, where service-industry employees scrounge at minimum wage jobs. There are no keys in the cars. All the trailer doors are locked. The Divide hides behind a blue carbon-monoxide haze. It sucks.

As evening settles in, I settle back in my truck and head south and east, out of the valley. On top of Berthoud Pass I stop to let this tired old truck rest and to take a look out over the darkening valley behind me. A few car lights trace their way along the contours of the mountain, heading up the pass. The only other lights are natural, the night's first stars piercing the turquoise sky.

7

THE GREAT KIVA

*T*HERE IS A STRETCH OF RIVER THAT RUNS BEHIND OUR LOCAL K-MART. AS WE FLOAT COMMERCIAL RIVER TRIPS within sight of it, I point and tell my customers that it is the Great Kiva of our culture.

They think I'm joking.

8

DRINKING AND DRIVING

I LIKE TO DRINK AND DRIVE.

This is, of course, a very uncool thing to do. And these days it's an even uncooler thing to say. But the fact remains: I drink and drive, and I like it.

Don't get me wrong. I don't mean that I get drunk when I drive; that's not the objective of my drinking and driving, and it's not the point of what I want to say here. What I mean to say is that for me there is something about driving along backwoods roads, like, say, a one-lane dirt road that wanders up a little valley and off into some distant brown mountains, while in my truck I hold the wheel in one hand and a beer in another, occasionally sipping while listening to bluegrass or old country and western or just the sound of the tires grinding gravel, with the window open and my elbow hanging out and the breeze blowing in.

There's something about that.

And what I want to say is, that something is more than what it seems on the surface.

On the surface, this backcountry drinking and driving is a rather silly, perhaps ridiculous, even stupid activity to praise. It may seem that the whole scene—a guy in a beat-up old pickup drinking beer and driving through the hills to the twangs of country music—is one to stick into the "redneck" file, an activity low in intellect, lower in social status, and devoid of benefit.

Maybe it is.

Still, for me it is something else.

For example: On a cool fall night, I pick up my friend Bob at his place outside of town. In my old Jeep we head west, further away from town and over a winding blacktop toward the mountains. As we start to climb the first rolling foothills, the pavement ends and the road thins to one lane. Here we stop. I hop out of the cab and pull on my warm coat while I sneak a look up at the blue-bright stars. I grab two beers from the cooler behind the front seat, then climb back in and hand Bob one of the beers. We open them and toast, aluminum on aluminum. I slide a Marshall Tucker tape into the player (alas, children of the seventies, we are) and roll down my window. The coolness of this autumn evening rushes in and over me like water, and we continue on.

We drive for several miles through a dark canopy of Ponderosa pines and open scrub-oak stands until we come to a place in the road where we can look out over the town, filling the valley below like a reservoir of light, a pooled cluster of yellow stars. We pull off and stop here, and get out to study the view. Except for the concentration of light below us, the only other lights are the real stars overhead, pinpoints of brilliance. The Milky Way spills in a fat stream, splashing a million-some drops over the rest of the sky.

Between the starry dome overhead and the bright town below is a dark void we know is land, countryside, the wild places Bob and I love and have built our lives around and upon. Unmarked and invisible in the moonless night, we still know it well enough to point out the places. Over there are the Twilight Peaks, to the right of those is Mountain View Crest, ahead is Missionary Ridge and to the right of that are the foothills around the Animas River Valley, rolling away into the distance and merging with the mellower mesa and canyon country surrounding the San Juan River.

We look out over this darkness and fill in the visual vacuum with words broken by pondering silences, with the quiet companionship of a pair who live for this country and who understand each other for that reason. We lean on the truck and talk, reminiscing about past adventures together and present tasks and challenges in life. I offer him a cigar, and take one myself. We light them and continue our night study, our conversation under the stars, our savoring of the cool night and cold beers and the coal-black landscape. The cigars' glowing tips flare and swing like meteors.

It is that glowing tip of my cigar that reminds me of something, of sometime, of someone. The finger-curl of smoke carries my thoughts up and out and to a past time and another figure holding a cigar, the same cheap, sweet brand I taste tonight. My great-grandfather stands silent—square-jawed, big-nosed, thin-figured and thick-haired—while my father and his father talk. I am young, not yet a teen, and I sit and listen, staring at my great-grandfather. I am in awe of this man, this figure of 90-some years. He is the last Indian in our lineage, a visible and living genetic link to another time and world.

I wonder if it's his genes that lead me to hold this cigar tonight. Or maybe it's an unconscious imitation. Or maybe I just like a cigar now and then, too.

I take another draw, exhale the fumes, take a sip of the musky brew. There are more familiar things than just my great-grandfather in this scene here tonight, in me and Bob out here in the hills with beer and cigars and a steel guitar serenading in the background. How many other times have Bob and I (or another good friend and I) stood before some vast expanse of land, leaned on my truck with beers, driven around the countryside in my old Jeep or a pickup or a beaten and tired old sedan? A God-awful many times. More than I can remember. We did it more often in the days when we were single and living on the odd jobs we could find and the fish we could catch. We did it more before we were married and were striving toward careers and had houses to maintain.

But we are still sure to do it sometimes, occasionally, every now and then.

This scene tonight is familiar, comfortable. It seems like now, it seems like 10 years ago. And it seems like 25 years ago, in the mountains of New England with my father and his friends, as they drove around in his pickup and talked, listened to country music, and—of course—drank beer and stopped in places with views above and over white-steepled towns in New Hampshire, Vermont, and Massachusetts. I know this happened because when I reached a certain age, in my early teens, I was allowed to join my father and his compatriots on those ventures, and together we relished views of light-filled valleys and lightless forests and ridges and summits.

A silly thing for a father to be doing. A ridiculous, perhaps stupid activity for a father to bring his young son along on. But for me, it was something else, even then.

To be asked to join in on those weekend drives over New England backroads was for me a rite of passage. And it was treated as such by my father and his friends. When I was given a place in the middle the Ford pickup's bench seat, I knew, and they made sure I knew, I had reached an age, a maturity, a new phase in my life as a young man. I knew I was becoming adult, and that I had that day been given symbolic entrance—at least one stage of entrance—into the world of adult men. More rites of passage would come that I would have to earn: joining my father hunting, carrying my own bow on those hunting trips, and one day being offered a beer myself while we drove through the New England countryside. But this first rite-of-passage was perhaps more significant than those others. It was the first great honor bestowed by these elders, and the act was sacred.

Sacred? This was just a bunch of guys—machinists, car-penters, and factory workers—who on occasional weekends headed out into the countryside with a cooler of beer, a road map, a pair of binoculars and a box of 8-track tapes. They were just New England-style rednecks. Still, even though they wouldn't describe it as such, these weekend ventures were sacred. It was at these times they celebrated their friend-ships, their lives, their histories together, the places they lived in and loved. I know this; even as a teen-ager I knew it. On those ventures I could sense the appreciation in the discus-sions, the reverence behind the actions, the value of the time.

Like I do tonight.

So why do I keep doing this? Is it the genes? Is it just nos-talgia, habit, or emulation?

No, it is something else.

My father and his friends in their redneck way were prac-ticing a ceremony they had been repeating most of their lives, and that their fathers had practiced and passed on to them. And today, I practice that ceremony to tap the energy stored there during all those ceremonies that went before. I do it because as silly as this ceremony is, as simple and common as its tokens and gestures are—the beer, the music, the cigars, the drinking and driving—they are ours, they are the Wrights', they are mine, they are significant.

They are rituals. They are incantations of what is important and vital and treasured in our lives, in my life: landscape, friendship, the out-of-doors. In these actions, these tastes, these smells, and these sounds are stored the cumulative power and feelings of all the times I and the other men in my lineage have done this, and to repeat it is to tap that reservoir of power.

And this ritual taps a reservoir of power of not just of myself and my family, but of what it means to be human.

We humans are symbol-making animals; it is what we do. Rituals symbolize the non-material, the ethereal things that are real but intangible. They are the condensation nuclei around which vaporous ideas and ideals condense and become tangible, visible, alive and real. We, as humans, need rituals to rekindle our joy and our happiness, and to mark and renew our lives' non-material treasures. We need them, but we live today in a society that places most of its value on the material, on *things,* and we have lost many of our traditional and passed-on rituals that celebrate the pleasures that energize life, that truly bring joy, that imbue life with the crisp and crackling energy of being alive *this moment.*

But even if our society ignores those rituals, we still need them. So in recent years I find many of my friends seeking rituals for their lives. They believe our European, industrial, heaven-after-death, consumer-driven and TV-sedated culture is devoid of meaningful rituals—which it is—so they look to other cultures for rituals they can appropriate: sweat lodges, medicine bags, vision quests, pagan holidays, dreadlocks, meditation, colorful non-Western clothes and music and lifestyles.

These things work well in those other cultures, and they offer these friends of mine and many people today useful tools. But in order to appropriate those other cultures' tokens and gestures and symbols and rituals, my friends had to give up their history, their heritage, who they and their families were and are. They have to live in some netherland between the culture they were born into and the cultures they imitate but can never be part of.

Perhaps they over-think what a ritual is, thinking that in order to work, to be evocative and powerful and joyous and focusing, rituals must be dramatic, exotic, and unusual.

But I think perhaps a ritual is where you find it, where you make it. No matter how trivial they may seem on the surface, we imbue things with the power of ritual by using them meaningfully, respectfully, deliberately. Morning's first coffee. An old worn shirt. A tree in the yard. A 1970s pop song. Drinking and driving.

Tonight is another stitch in the old fabric of life for us Wrights, a fabric woven together by rituals. And tonight, here with Bob and the truck and the views, the whole simple and silly experience wears like a well-worn, familiar, and comfortable coat. Our coat.

9

MY GREATEST FEAR

*M*Y GREATEST FEAR? IT IS NOT THAT OUR TECHNOL-OGY WILL FAIL TO KEEP UP WITH THE MANY PROB-LEMS industrial civilization creates—pollution, ecological degradation, social inequality, the population explosion, food and water shortages, growing energy and resource hungers. My greatest fear is that technological solutions *will* work, allowing us to swarm the planet in ever greater numbers, to grow ever more estranged from our wild human-animal natures and from wild nature itself, to become hopelessly enslaved to our cult of economics, and to finally domesticate wilderness until the planet is one, big machine, serving humans alone.

10

THE VALLEY

*T*HERE IS A VALLEY, A LONG AND SLENDER FOLD IN THE EARTH PARTING THE FINGERS OF A RANGE OF PEAKS, THAT I visit several times each year. To reach this place, I fly up a four-lane interstate until I turn onto a state highway crawling over a low-point in the mountains. When I leave the interstate, I leave a steady flow of fast traffic, exchanging it for the slow swing of switchbacks and, after an hour or so, a measured roll through a couple of mountain towns.

I pass through the towns, then turn left, west, onto a dirt road. My little truck rumbles over gravel and past two National Forest campgrounds, where smoky fires sparkle next to tents, camper trailers, and brown pit toilets. After the second, and last, campground, the road immediately narrows, roughens, and bends, snuggling alongside a small creek. Most of the vehicles I encounter on this stretch of road belong to either fishermen or the Denver Water Board, the agency with "rights" to most of the water in this drainage. I pass several of their diversions, green-steel and concrete barriers herding the creek east, into ditches and tunnels and eventually under the Divide to the East Slope.

A couple more miles and I pass the last diversion, and

again the road takes another drop in quality, gaining personality. I double-clutch into first gear, lurching through washouts and over exposed bedrock. Opening my window and leaning out, I look down onto the creek, entrenched in a deep notch, rolling noisily over rocks and deadfalls. The air is cool. It is always cool here, perfumed with the sweat of Douglas firs and Engelmann spruce. Looking up, through the forest greenery I see flashes of bare, sky-scraping tundra.

My truck lumbers along loudly over the steep and rocky last few miles, until the road ends abruptly, as though the bulldozers went home on a Friday night and never came back. The reason for the sudden terminus is clear enough once I shoulder my pack and start up the trail, which continues from the end of the road toward a high pass in the distance. Immediately the valley pinches into a slot of granite bedrock, cut into a V-shape by the creek. Pushing the road farther would have been too much work for too little reward. The good diversions, good campsites, and good fishing are all downstream.

And so was the good timber, once. Up the trail I soon encounter fat, crooked old-timers of trees standing tall and solid, mixing with younger, frailer kin. Strewn erratically across the sloped forest floor lie deadfalls, ancient behemoths felled by lightning and age. Stands of thick shrubbery and pine seedlings hoard the sunshine offered through these openings in the canopy. In these last few hundred yards before treeline, the forest stands as it has for centuries, as it once stood throughout the valley. In the last hundred years, all but these upper nooks of the valley's tributaries has been logged. These remnants survived the lightning-strike of economics by being blessed with harsh terrain. Like the road, too much work for too little reward.

Two hours later I am sitting on a shaft of granite that juts out into the bowl-shaped headwaters of the valley. I am encircled by open space. What does that mean? In the distance, nearly ten miles away, squats a small group of buildings—the town where I picked up the dirt road—huddled on the valley floor and humbled by the immense backdrop of the ragged Divide and its rolling foothills. The town sits at the junction of the valley before me and a still grander and greater valley, a land of dreamlike beauty, a high-mountain-scape of meadows and marshes and forests and creeks. I suck in this vision and exhale an overwhelming feeling of good fortune.

And where, exactly, is this special place? It doesn't matter. Its name, its longitude and latitude, the route number of the highway that accesses it don't matter. Although I have a singular relationship with this valley, this is not what I want to celebrate here. This vast landscape before me is a symbol, an example, a representative member of a larger tribe of landscapes: This valley embodies the allure of the entire West Slope of the Rockies, my home, a domain rich in places at least as magnetic as this one, yet each distinct in character.

Geography blesses the West Slope liberally with broad mountain vistas, open-roofed expanses of deserts and plains, rampart-rimmed troughs of valleys and canyons. The fact that these places, like the one I peer over from this rocky perch, comprise the least-trammeled chunk of our over-industrialized continent owes itself to other blessings—curses to industrial developers—which, so far, shelter this place from late-20th century economics: rugged terrain; little water; nutrient-thin soils; lots of rock, ice, dirt, and alkaloids; remoteness from urban centers; federal ownership of much of the land area; and seasonal extremes of heat, cold, snow, thunderstorm, drought, flood, and wind. Trying to make money here (except for hit-and-run raids on resources) is too much work for too little reward. Thank goodness.

To be sure I'm not misunderstood, let me say that I'm not a misanthrope. I like people (some), and I need people (sometimes). But I like my people the same way I like my land: humble and relaxed, inviting but quiet, open but mysterious, rough yet inspiring, mostly undeveloped, and, importantly, spread out. Like the West Slope. Like the people who are drawn to the West Slope.

These characters who endure and relish the challenges that living on this landscape presents become themselves a stitch in the fabric of this place. Surviving here is not easy; you must carve your own niche. Work is scarce, low-paying, and often seasonal. Boom-bust cycles are common. Without big industry to provide steady and stable business, most work is somehow tied to the land, directly or indirectly, in the form of agriculture, tourism services, or resource extraction like mining or logging. Even day-to-day living is demanding, whether it's starting a car in 30-below-zero weather, carrying clothes for a hot day followed by a frigid night, or driving 60 miles of snow-packed road for milk and bread.

Why do we do it? Why work so much for so little economic reward? Because we can sit alone on a rock like this, overlook a valley like this, and hear only the distant thudding of a woodpecker luring insects from a tree; the lulling hiss of cascading, converging, snow-fed streams; and the music of wind drawn across the strings of trees. And we can return, when we want to, when we need to, to our little villages—Paonia, Bedrock, Paradox, Bond, Tabernash, Dolores, Durango—and not have to explain that look of contentment on our sun-baked faces that in the city might draw suspicions.

I recently took a friend visiting from the East over the Divide. I wanted to share with him what I had found here. I wanted to take him from his world of stores and industry and computers and commutes and subdivisions and pavement and have him experience the wildness of the West Slope. I wanted him to gain this understanding through the valleys and towns and people, as I had.

I was very excited and drove him around some towns, showed him the meadows dividing them, visited the liquor store and tackle shop, the only businesses in the town I was living in (what else do you need?). We gazed over the craggy spires of mountain ranges, drank beers with locals at taverns. But he didn't get it. Things are far apart, he moaned. What do you do on weekends? he wondered. Groceries are pretty expensive, he observed. Aren't these roads hard on your truck? he asked. And lastly, around midnight, he began planning: Just think what you could do with some money up here, he mused. Just think!

But at that point I didn't want to think, especially about what my friend saw with his Eastern eyes. I was disappointed and a little drunk, and in response I was overcome by the urge to sleep out-of-doors. My friend was game, even though it was February and snowing furiously. I had camping gear in my truck, so we wound our way out into the forest and bedded down in the blizzard.

In the morning we awoke covered with snow but warm and chipper. Overnight the storm had cleared. I fired up a pot of coffee, and with steaming mugs in hand we stood in the midst of the muffled silence of the carpeted woods and stared out at the white slopes of the Divide, glistening like God's own jewelry. My friend just looked at me and smiled. He understood.

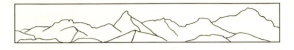

11

A Modest Proposal
For The Interior West

People have the power
to dream
to rule
to wrestle the earth from fools.

—Patti Smith

I LIVE IN A BROAD, COLORFUL VALLEY OF STUNNING BEAUTY. IT IS LINED WITH FANTASTIC RED AND WHITE bands of rock that roll back to forested foothills and, further back, a series of jagged mountain ranges. In the valley's wide, flat flood plain meanders a river. It is slow here, taking its time around bow-tie turns, occasionally calving an oxbow lake, a cut-off meander that turns into a moon-shaped pond, a wetland, a watering hole for elk, deer, and cattle.

In this valley the river rests after a mad plunge from the mountains, where the river is born from hard winters and lingering snowfields. It doesn't rest for long, though; after sliding through this particular piece of valley it picks up the pace again, straightening, narrowing, and entering a new incarnation, this time a desert river in canyon country, its final and longest life-phase that will (or at least once did, before the Age of Dams) take the mountain waters all the way to the sea, and to the death and rebirth that is the ocean.

But that is far away; it is the valley I am concerned with here. In this valley is a town, where I live with my wife and young son. The town itself straddles the river at the down-stream end of the valley, just before the river wanders off into its first desert canyon. The town fits into and is shaped by the valley like it was poured into a mold, with some overflow spilling up onto a couple of nearby mesa tops and side drainages. Upstream of town, the population thins to valley-side homes and, along the flood plain itself, ranches with patch-work meadows and stands of cottonwoods.

The town is relatively new to this valley, having officially incorporated a little over a hundred years ago. Still, people have been sinking personal, family, and community roots here for at least 2,000 years, that we know of. Evidence lies in caves and on terraced slopes where archaeologists are slowly uncovering the pit houses and early kivas of a number of ancient communities. These people farmed here—some of the earliest evidence of farming in the Southwest—and they created art, raised families, hunted, and explored.

They lived good lives, I suspect, even though theirs was far different from what we call a good life. I have been to some of these ancient sites, and most have expansive views of cliff faces, river bends, and elk and deer (and, probably then, also buffalo) pasture. I don't believe this was an afterthought, or can be attributed solely to the practical rationale of defense or food supply or solar exposure. I think these people *liked* the views. Of course, the only evidence I have to support this theory is that I live here, too, in the same valley, and I like the views.

Why do I think that my living here allows me to project these feelings onto past residents? Because I believe that I and most of my contemporaries live here for fundamentally the same reasons that people have lived here for 2,000 years or more. Over this time the appearances of the people and their lifestyles may change, but these surface structures are built on the same foundation: the landscape.

Take, for example, the present incarnation of human habitation, our little town. Today, it hums along much as it has for more than a hundred years, although these days there is a higher ratio of professionals to the ranchers, loggers, and miners who dominated the town in the past; and there are more folks who get on the land to recreate rather than to create income. For the most part, though, the residents here still settle and linger for the beauty of the valley, the wide-open expanse of surrounding desert, the wild mountain country, and the remoteness of this place from the general human mass of the rest of the country.

This doesn't apply to everyone, of course. There are some who live here because the land also offers jobs, material wealth, and a chance to make it. There is the energy industry and its roughnecks, the water engineers and resort developers, and the ubiquitous real-estate speculators. But those opportunities are limited, and most of these people will move on when they have built their projects ,or the mother lode or gas field is played out, as they always have. They will seek the next big strike, the next ski area to be carved, the next ranch to be subdivided, or they will retire where they can spend their last days putting balls and sipping gin and tonics. These people didn't come to this valley for the land itself, so it doesn't anchor them; they came for what the land offered in opportunities for personal gain.

For most people seeking personal fortunes—or just looking to get by, for that matter—there are a lot of easier places to make it than in this remote valley. In most any American urban area there are more jobs and opportunities for advancement than here, and housing, amenities, and entertainment are more available and affordable. Compared to most of the country, getting by here takes a lot of work and risk and enterprise. Why? Because even in the late 20th Century this valley remains relatively isolated and remote, and access is relatively difficult. We get by with only a tiny airport. Driving here means hours on long, slow, winding two-lane roads, most over difficult, frightening, often dangerous mountain passes. And we're more than just physically isolated: We are in a distant eddy, a social oxbow, a cutoff meander of the main flow of American society and culture, of music and movies and fashion and current events.

For some of us residents, though, that's just how we like it. That's why we live here. We accept, even relish, even desire the work, the challenge, the risk, the isolation, and the slow cultural pace. We're not here to find the easy life or to make a lot of money; we're here for the place. So although there may be a seeming diversity of people living here, working a wide range of jobs—ranching, ski patrolling, newspaper reporting, banking, river guiding, cooking or waitering, farming, retailing, and teaching—there is one binding similarity: We want to be here. We seek this valley's non-monetary wealth: its beauty, solitude, wilderness, stability, and community.

But things are changing.

◆ ◆ ◆ ◆

In the last few years there has materialized in our valley three golf courses, hundreds of massive houses, and several suburban-style "planned communities." Dozens of large ranches have broken into thousands of ranchettes. Hidden in the forests around town are exclusive developments surrounded by elaborate fences with guarded entrance gates. A Wal-Mart, a third major grocery store chain, a five-plex movie theater are built or are building. The airport is enlarging to accommodate jets, and the mountain pass roads are getting straightened and widened. The ski area is expanding further into the national forest, and condominiums are under construction nearby by multi-national resort corporations.

Downtown is transforming into ethnic restaurants, espresso shops, outlet stores, and expensive hotels. To feed this growth, a massive federal water project intends to drain the river and flood a basin vital to elk winter range.

And people are moving here in swarms.

Why this recent eruption of development and population? Lots of theories are being tossed out: People are fleeing the urban nightmares of traffic and crime and pollution, and the suburban prisons of blandness and boredom and mediocrity; the baby boom is going through a mid-life crisis and is looking for a collective change in its surroundings and routine lives; older folks are retiring to the country where their pensions and home-sale money go a lot further. All these rationales are supportable and probably at least partly true, but there is one rationale notably absent: No one really thinks that suddenly a huge chunk of the population became enlightened, is now willing to sacrifice their jobs and urban social and cultural amenities to live close to some fantastic landscape, to challenge themselves to live simpler, more spiritually rewarding lives, and to join and blend with the rural culture that has lived this way for a long time.

So why are so many people moving here? I offer a simple explanation: because now they can. Decades of growth promotion and infrastructure improvements in this valley are forging a new human habitat, one that is readily accessible to and from the rest of the world, offers more jobs, is built in the style of American urban and suburban environments, and offers timely and popular cultural and social activities. Now, urbanites and suburbanites can dump their old lives and transplant here without really giving up the ease, stimulus, securities, and occupations of the lives they knew before. These people may like the landscape here, but they don't have to sacrifice for it, earn it, accept it on its terms. The mystery and beauty and allure of the land are not their highest priority; those are just the proverbial frosting on the cake.

And with all these new folks sitting down to eat, that cake is getting gobbled up.

✦ ✦ ✦ ✦

So who's to blame for this predicament? A lot of people who agree with what I have said so far blame the newcomers. It gets ugly around here. Newcomers get called names on the street, get railed against in the newspaper, have their houses and cars vandalized. To drive around with California plates (statistics show most of the new arrivals to our valley are from California) is to invite rude gestures, name-calling, or vandalism.

But, really, whose fault is it that this once-isolated valley—and the entire interior West, for that matter, for what is happening in this valley is happening all over the West—is being invaded? It's not the fault of the Californians and Illinoians and New Yorkers who are moving here; it's Westerners' fault.

For decades the West's political and business leaders have promoted and advertised and built the West. While rural Westerners have perhaps not cheered our leaders on, they did not stop them, did not take control, and in fact accepted the jobs boosterism and growth offered. Now the devil has come for payment on that Faustian bargain, and he drives a Ford Explorer with California plates. So today we find the once-rural West trapped in a perpetual-motion machine: People are moving here in droves, and they are bringing with them their urban needs and expectations, and these people's voting and buying power fuel ever more urban/suburban-style development, thus luring more of these people.

And where is this headed? Soon this place will be just like every other place. Prettier, maybe, with its mountains and valleys and deserts, but socially and culturally just the same, as easy to live in as Newark or San Diego or Manchester, New Hampshire. Falling in love with the American West is like marrying someone with a terminal illness—the love is real and passionate and enriching, but you know the end is coming fast, too fast, and you have to watch the disease eat away at your lover's health and beauty. And by proxy, it eats at your soul just as much.

Still, this illness continues to be praised by our politicians, business leaders, chambers of commerce, engineers, investors, land speculators, realtors, etc. By these people's standards, the growth sweeping the West is good, natural, progressive, and beneficial (as it certainly is, to them). These changes are the textbook definition of the "growth" that fuels the economy, supports the nation, improves our collective standard of living, multiplies our Gross Domestic Product. But some of us don't want that; some of us prefer quality of life over quantity of things. What do we value? Healthy landscapes and communities; economies that are sustainable over the long term, and lifestyles that are stable over generations; and an environment that nurtures an individual's quest for joy and personal satisfaction. The growth and change sweeping the rural West is also sweeping away the opportunities for that quality of life.

So where's the rebellion? Where are the Sagebrush Patriots fighting off the invasion of our land and our communities and tar and feathering our traitorous leaders? The emotions are there; I hear the pissing and moaning all over, at work, at the local pub, at the grocery store, in the letters to the editor. But most Westerners seem to be content to act as pall bearers, singing the mournful choruses to the West's funeral dirge: *There's nothing to be done; change is inevitable in this modern world; growth is necessary for a healthy economy; you can't tell people what they can and cannot do; the economy is going global and the West will have to join in; and so on.*

The arguments for not rebelling against the growth overrunning the West are tossed out with strong emotion, sound reasoning, and good intentions—when you start from a particular point of view, that is. These are good arguments if you assume that everybody wants or needs ever-increasing material goods and incomes, an easy and efficient lifestyle, access to the global industry and technology system, an economy that needs to grow and grow and grow ad infinitum. Doing nothing but bitching makes sense if you assume there's no way to stop the economic and population tidal wave sweeping over us.

But what if we change those assumptions? What if we place as our highest priority needs that are non-material, non-industrial, non-technical, non-growth-driven? What if we seek truly human needs? What if we really wanted to preserve small, close-knit communities set in big, open landscapes? What if we wanted a sustainable economy based on small-scale farming and ranching and other land-based lifestyles? What if we wanted the West to continue to be a place where those values could be lived, now and for generations to come? What would we need to do?

I have a proposal: Keep the interior West a challenging place to live. By preserving the American West as what it has been since Europeans first began developing this continent— the wildest, remotest, least developed, and least accessible corner of the country—we can create a refuge where people can work on growing the human spirit rather than the global economy, where people can seek a higher quality of life rather than an ever-higher standard of living, and where people can rest assured that their children's children will know the same sense of community and place as they know.

Let me make a few suggestions:

Keep access to the interior West challenging. Access is a natural filter for both quantity and quality of people in a place. Most people don't like to work too hard to get somewhere, can't take too much time, and don't want to take too many risks in their travels. So between the Sierra Nevada and the Great Plains, roads should be simple—adequate, say, for two-wheel drives in good weather.

Driving into the interior West should be a serious undertaking, a decision to be pondered over, planned for, and assessed carefully. Except for jaunts to jumping-off towns like Reno, Denver, or Bend, Oregon, trips in the heart of the region should not be toss-the-kids-in-the-mini-van-and-go trips. These excursions should be safaris, physical and spiritual adventures with the attendant risks real adventures entail, such as getting lost, stranded, broken down, or worse.

Why not fly? The exorbitant airfares airlines charge are built on the fact that people will pay a lot to get somewhere quickly and easily. Flying makes life in even the remotest of places just a hop from anywhere else. To discourage the obliteration of space and effort that airlines encourage, airports in the interior West should be dirt airfields, scratched into the mesas and meadows around a few of the bigger towns. Bush planes and small operators can make a go at scurrying around the desperate and the brave, but flying will be a limited and adrenaline-pumping option, at best.

For people living in the heart of the West, isolation should be a fact of life. Getting around beyond the immediate community will take time and energy. Vacations to the outside world will be rare and memorable events. The result: The interior West will be a region isolated and remote by choice. To the residents, the loss of accessibility will mean a gain in independence and self-reliance, and a physical and psychological commitment to residency. The resident population will be pared down to the folks who really want to be here. As for the people who choose to visit despite these challenges, they will find, right in their own country and within no more than two days' drive of either coast, not a place for a casual visit—there are plenty of those elsewhere—but a vast area to explore and adventure in, big enough to spend weeks crossing if one is so hearty. And these travelers will go home with tales of the big outside rather than just more made-in-Taiwan coffee mugs painted with scenic pictures. And those who don't like the poor access will still have plenty of other beautiful and convenient resorts to visit, or paved and interstated places to live.

Reduce to a limited infrastructure. Like access, the availability of utilities is a filter for the types and scale of industry, businesses, and residents in a region. For small-scale, community-based, human-powered industries that employ and meet the needs of the resident population, an early 20th-century infrastructure should be sufficient while still offering residents some of today's simple luxuries such as electricity, running water, and central sewage. Small water-supply reservoirs and stock ponds can replace massive and expensive dams. Big power plants, pipelines, and other modern mega-projects can be substituted with cooperative projects that meet each community's minimal and essential needs for power, water, and public works.

Without big utility projects funded by federal taxes, big business, and the exporting of resources, residents will have to evaluate the real costs of their needs in terms of capital, labor, and damage to the surrounding countryside. The environmental limits to water and energy supplies will have to be recognized, and those resources conserved. A lot of water-and-energy-sucking enterprises will be lost. For example, Las Vegas, a micro-pseudo-culture built on pumped water and electricity will lose it underpinnings, as will the Bureau of Reclamation, a micro-pseudo-culture built on pumping water and electricity. Irrigation-project-based agribusinesses, coal strip-mining companies, cyanide-leaching gold mining operations, and the intermountain West's natural gas grid will all have to either pull stakes or continue their work honestly, with hand tools. Computer firms, banking enterprises, and grandiose all-season resorts with world-class accommodations, gondolas, golf courses, and marinas on man-made "lakes" will have to relocate to locations better suited to their endeavors than is the arid West.

With this reduction of energy and water demands will

come the added benefit of a cleaner, healthier landscape. The West will be rid of its river-strangling dams, freed from its air-clouding and lung-corroding coal-fired power plants, relieved of its landscape-trashing mineral extraction operations, unleashed from its horizon-cutting power lines, and spared its scarring oil and gas pumping system. The rest of the country that relies on the plundering of the interior West for its energy will have to look elsewhere, or learn to use less.

Declare the interior West a refuge for individual and human-scale free enterprise. I mean true free enterprise—local individuals working locally—not the corporate capitalism that is pawned off on us as free enterprise today. To do that the West needs a self-contained economy where regional human and natural resources serve regional needs, and where national and multi-national corporations can't undercut the prices of small local businesses or compete with resident entrepreneurs for the region's resources.

Locally owned industries that service a local area have a stake in the welfare of the community and long-term health of the land, and they must be responsive to the community's best interests. Yet these locally-linked businesses cannot compete with the economies of scale that allow big corporations to operate and offer products and services more cheaply. The interior West will have to serve, service, feed and supply itself through individually-owned and operated businesses that run on people rather than machines, and that are based in the region's communities rather than distant cities and countries.

Isn't this an inefficient, archaic, un-economic, backward, hillbilly way of doing things? Yep. It is also personable, simple, labor intensive, egalitarian, sustainable, and empowering (to use a popular term). Won't it also be more expensive? Perhaps. Probably. There's little question that the global marketplace works to give us a huge variety of inexpensive things. But what are we to value more: A broad variety of inexpensive furniture (for example) from a Georgia-based company that clear-cuts in Idaho, mills the lumber in Mexico, and builds the table in Guatemala, or a smaller selection of furniture built from wood cut on that hill over there, that we buy from the guy who employs several people in town and who we went to high school with, and whose kid (who is learning the trade and will inherit the business) hangs out with our kid? Which piece of furniture is cheaper depends how you define costs. In a contained and localized market, the true costs of a product cannot be hidden in the destruction of remote land; dispersed air, water, and soil pollution; or the exploiting of foreign labor at the expense of local jobs. In the local marketplace, owners, laborers, and clientele will all eat from the same trough and belly up to the same bar.

And without those massive corporations owning and streamlining and running the industrial show, there will be the opportunity for a lot of producers to go into the business of meeting the region's demands. Niches will be abundant for individuals to find work at everything from laborer to craftsman to entrepreneur. This will be especially true for agricultural products, where the family farm and ranch will make a comeback meeting the food needs of both the owner-families and the community. Bartering for goods and services will displace much of the cash economy, clearing the way for still more cash-poor people to enter the region's economy.

Protect the West's public lands, both ecologically and politically. More than three-quarters of the interior West is public land, and this is its greatest, most unique, and most vital resource. Few other places on earth have such a store of open land, wilderness, and wildlife habitat held in reserve for the future and the common good. That resource must be defended and preserved. How? No more use of public lands by big corporations, and no more exporting of natural resources. No longer will a company from Australia strip mine coal from public land for shipment to Japan, or a Canadian company be able to buy public land rich in gold for $2.50 an acre. Public lands should be used by local people for individual needs such as meat, wild foods, fuel wood, and raw materials, and for locally owned enterprises that are small-scale, low-impact, and that the land can support in perpetuity, such as pick-and-shovel mining, hand-saw and mule-team logging, small family-owned livestock operations, and wilderness hunting and safari guiding services.

Public lands must also be managed for non-human life by protecting ecosystems for future generations, which will inherit a living landscape that they are part of, a wilderness that includes humans and their uses—true wilderness. This will also open a unique business opportunity for the residents of the interior West. As wilderness is increasingly lost in the rest of the world, and as the little islands of protected wilderness there get increasingly crowded, people from outside the interior West will want to sample this region's vast reserve of wild places. They will come seeking some sanity, some taste of the natural world, some sense of what rural life is like. Guide services will flourish, and without the dominance of corporate resorts, family owned bed-and-breakfasts and restaurants will thrive. Of course these visitors will have to meet this place on the land's and the communities' terms—wild, difficult to access, unrefined and without modern conveniences—but this will be this region's allure; people from the outside will come to step into a real world of human community and living.

Again, for those who can't leave their Winnebagos or Monday Night Football or fax machines behind, there is still that abundance of glamorous resorts and industrial-tourism destinations elsewhere.

Lastly, don't encourage people to move here. Give tax breaks for conservation of land; no tax breaks for second homes. Do no promoting, no boosterism, no advertising. Abolish tourism bureaus. Disband all chambers of commerce. Word will get out to those who care.

✦ ✦ ✦ ✦

Those are just a few ideas to get started. A modest proposal.

Believe it or not, there are going to be some complaints about these suggestions. *My God!* people will gasp, *how can you tell people they can't have all the electricity and water and computers and subdivisions and high-paying corporate jobs they want?* They can have them, just not here. People who want that stuff can live anywhere else in the world—which is most of the developed world—where modern life rules.

But, others will cry, *Westerners won't be able to compete in a global market economy!* True, true. But that's the point: Not everyone wants to be plugged into the global frenzy. The American West will be a hold-out, a refuge, for those people who want to focus their energies on living, on the aesthetics of working.

This infringes on Americans' freedom! others will scream. But this isn't about freedom, this is about rights—not property rights or Constitutional rights (which are essentially the same thing) but the natural rights of human beings: the right for every person to carve his or her own way, to live simply and affordably in a secure community, and the right for free access to abundant and healthy wild land. And it is about responsibility—the responsibility of our generation to guarantee those rights to future generations. Where is your freedom when you're getting bulldozed for development and stampeded by population? When economics makes it best for you to trade your land and home and community for dollars? When you lose your job to a computer or machine? All long-time Westerners have the freedom to do now is cash in, jump on the development bandwagon, and profit from losing the unique and irreplaceable place and lifestyle that is rural life. Once that is done, there's no going back. Remember: Development is Forever.

Then there's the "elitist" charge. *Isn't this selfish for the relatively few people who would want to actually live in this primitive way?* Perhaps. But all people will have the choice of living here, they just will have to do it on the land's and lifestyle's terms. If you are unfit or unwilling to live this way, then, as I said before, there is a rich and abundant diversity of fully developed places to dwell. The people living here will be those who want to live here, people who live on small incomes either by choice or because they lack the marketable skills demanded by a high-tech, global-marketplace economy. I mean the crafters and tradesmen who work with their hands, and the small-scale farmers and ranchers who work in the out-of-doors—people who survive mainly in cigarette and truck ads, or in catalogues featuring expensive rustic clothing. I also mean the unskilled, the uneducated, and the people who voluntarily reject the 40-hour-a-week-with-two-weeks-off-each-year indentured servitude we are offered— people who are erased from the mass media altogether. These are the same people who are increasingly pushed off the land and out of communities by the growth transforming and deforming rural places in the West and elsewhere today.

A modest proposal: Let's set aside a little piece of our industrialized, urbanized, globalized, and sanitized country for *people,* for people who want to live like real humans doing real work in a real landscape. Let's offer a choice: Let's leave some of the country reserved for those who believe in, want, and need a rural lifestyle, wild country, and wild lives.

And who are those people, exactly? There is no "exactly"; we bear many signs: Cowboy hats, camouflage caps, and baseball caps worn backwards; Lycra, blue jeans, shorts, and baggy tie-dyed pantaloons; ski racks, gun racks, and bike racks; Wise Use, Earth First!, and the Sierra Club; fourth-generation ranchers, college students, and, yes, even some of those new-comers from the coasts and the cities who are setting up camp in the rural West. We are not easily stereotyped; we are everyone who came West for the land, for the communities, for the life. We are all the people talking about leading quality lives in places we love, passing safe communities and healthy land onto future generations, and the right and responsibility of the people who live in those communities to shape those things for themselves.

And when we recognize each other, when we see who our true allies are and who the real friends of the interior West are, we will see that we are a force. A force big and strong enough to take charge. It's not too late, if we just start dismantling....

12

WHAT MATTERS

*A*CTIONS MATTER.

Words matter,
joy matters.
awareness matters,
landscape matters.
responsibility matters,
spirit matters,
truth and honesty and honor matter.

Actions matter.

Symbols, ideas, and myths matter,
community matters,
family matters,
play matters,
diversity matters,
character matters,
conscience matters,
rituals and gestures and celebrations matter.

Actions matter.

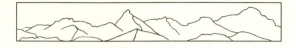

13

WILD CHILD:
WHY THE ENVIRONMENTAL MOVEMENT
NEEDS PARENTS AND CHILDREN

Her womb is the first landscape we inhabit.

—Terry Tempest Williams

*T*O BREED OR NOT TO BREED, THAT IS THE QUESTION.

That is the question, at least, that haunts some people I know. It haunts those who defend wild places, who fight for a clean and living world, who resist our insatiable society's consumption of resources. It haunts those who see our late-20th century world as a still-filling cup already brimming with humanity.

The question doesn't haunt only friends of mine. In periodicals, at meetings and conferences, in casual conversations wherever I encounter environmental activists around my age—breeding age—the question hangs in the air like a guillotine blade whenever the subject of child bearing and rearing comes up. To breed or not to breed? Can people fighting to keep the world from turning into one big L.A. justify having children themselves?

I've pondered this. And while I was pondering, my wife and I... well... had a kid. And after a year of thinking with this little person in my life, I have an answer for all you thoughtful and ecologically ethical people who question the righteousness of environmentalists bringing more people into the world. Yes, I answer that quandary, some of us in the environmental movement should, even must, bring children into the world.

Let me explain.

I don't think I'm a slacker. I have worked to defend wildness and wilderness, and some of these actions can be considered radical. I have educated myself about issues and natural science and strategies of activism, and I have applied and shared and taught what I have learned. I have been a college instructor, a wilderness guide, a park service ranger. I have walked and boated and lived in a tent. I have written letters and news stories, blockaded stores and offices, been on T.V. and broken the law. I have dedicated my life to experiencing wilderness, and to seeking a perspective on and acting on what it means to be a human who is part of the web of life.

Then Webb came to life. A son, deliberately added to my home and to my world, our planet. And after a year with this child by my side, I can now say that becoming a parent is the wildest and most radical thing I have ever done.

In the last year I have witnessed my boy's little curled, gasping, blue body emerge from the core of my wife's body. I have watched him nourish himself and grow from what he can suck from her breasts. I have seen him caress his own hands, opening and closing them like hinges while he stared at them big-eyed, mouth agape. I have observed him at a month old pick out birds in flight with his eyes, at seven months point at birds at a window feeder, and at a year calmly state "dee" when a red finch landed on a branch over him while he sat in the grass.

And what do these and hundreds of other first-year-of-human-life events tell me? I am a mammal. I knew that before, rationally, thoughtfully, but until now, until I watched this growing, learning, scurrying little critter, this little animal borne of my self and my wife, I never really felt it. I thought I had, but now I am aware on a deeply physical, beyond-the-rational level. I, Webb, you, we are all earth-borne animals, from the deepest corner of our kidneys, to the roof of our psyches, and through the stratosphere of our spirits. Giving birth, producing milk and offering food, sheltering and protecting through infant helplessness, nurturing and sharing and expanding the world for an evolving awareness. Mammalian life, reproducing and eating and learning and growing and living on this earth like all other animals. That is all we are. That is more than enough.

Against this awareness, all the rest—politics and culture and economics—suddenly stands starkly contrived, made up, crafted, fictitious. Pulled by a tiny hand I have been led down a new trail, to new terrain where I am more bonded and committed, a more dedicated and hopeful and determined defender of a wild earth than I ever dreamed possible. I have found a source of energy and vision that stretches beyond what I can see or even conceive of, reaching into both our genetic past and our generational future. I am still immersed in politics and culture and economics, strategically, but I now live richly aware of the real world of our animal humanity.

Our clasped hands, mine and this little person's, are a link in a genetic chain, a multi-millennia-long line tying together hunters and gatherers with the seventh generation to come. We are all the same; we are inseparable. With Webb as my guide, I draw on the awareness of the former as I work for the world of the latter where there is still wilderness, wild creatures, wild rivers and wild country where wild people live wild lifestyles.

With Webb by my side, I can never consider despairing in or writing off the future of humankind. I have put my genes where my mouth is; I now have a genetic investment in the future, in the planet and humanity's life on it. Having a kid is the ultimate act of faith and hope.

But so what? So having a kid has deeply affected me. As good as it might be for me personally, this would still be mere selfishness if it weren't good for my place and my community as a whole, and if it weren't a vital step toward long-term change. I believe it is both. I believe that environmentalists who become parents regenerate the environmental movement. I believe that if we all—parenting and non-parenting activists—recognize the value of bringing children into our fold, if we all participate in their educations and their experiences, if we all feel a part of their families and included them in our communities, our networks, our tribes, then the environmental movement would transform from a mere movement to something more lasting, more effective, more real.

Take Webb, for example. What will he do when he wanders off down that genetic trail into the wilderness that is The Future? I hope he will join the ranks of a generation of children born appreciating, celebrating, and defending the wild world. Backed by their lifetime of experience and upbringing, this next generation will stand as the vanguard, form the bedrock of whatever the movement builds in future decades. I could be wrong about this, for I am constantly reminded that raising children is an inexact science, but that is the risk we take.

And that is what we must work toward, and to give them their best chance, we—both parenting and non-parenting environmentalists—must together raise this next generation. They must be seeded, grown, and nurtured from the soil of today's activists. But I don't see this happening. Instead, today I see the eco-warriors who live their strategic, lean, light lives piling guilt on the family people who are just as committed, but also are just-as-strategically settled and nested; and I hear the community-based activists laughing and pointing at the free-roaming, self-exiled outside agitators and alternative life-stylers. Where is this headed? Division, distrust, and distraction. We must be diversified in our strategies, but specialization of eco-strategies into isolated islands of dogma leads to a fractured movement that is easily divided and conquered. And in the future, as this generation of environmentalists ages and dies off, I see a movement—or several smaller spun-off movements—that must rely on conversion of outsiders to carry on the work and to lead.

There is another way, though. What if we bring together environmentalist parents and their kids with those who for reasons of conscience or strategy decide not to have kids? What happens when all of us moving the movement combine our efforts to nurture an enlightened generation rich in experience and ideals? If those things happen we begin to evolve more than a mere movement; we form community. More than that: we create a new *society*, a natural congregation of people encompassing the full spectrum of the human experience, people filling their best-suited roles—warriors and philosophers and parents and community builders— teaching and learning and working and fighting together.

Not everyone should go out and have kids. Parenting is the best path for some; good parenting is their skill, their predilection. These people, if they decide to have a child or children, should do it deliberately, carefully, well. But those who choose not to have kids still must understand the value of kids, and they need to support those in the environmental movement who do have kids, they need to see the opportunity these kids represent.

I don't mean we should pump these kids with propaganda, but the full range of the environmental movement's members can offer kids a spectrum of involvement, experience, example, education and learning as principles and skills; can show openness and communication and awareness of the world;

can provide diversity, community, ritual and celebration. I also don't mean to indoctrinate or force participation on kids, but we can offer them a sense of belonging and membership in active environmentalism. We can show them we recognize their presence and value, and offer a readiness to be there to answer questions and to share ideas and knowledge.

Backed by the variety of experience the full movement can offer, these kids will one day make their own choices, create their own movement. Or perhaps they will forge something more. A movement is a product of milieu, a reaction to immediate circumstances, but enter the children and you transcend the present time and reach forward, to the future, and become a lasting, resilient, adapting, evolving society.

That is what we need to do to survive, to effect real, long-term change, to turn ideals into reality.

We need the kids.

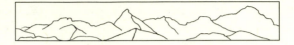

14

SARAH

FOR MY WIFE, WHOSE BOND WITH THE LAND AND WITH
CHILDREN IS A SOURCE OF ENDLESS AWE FOR ME.

Rolling in a breeze
   the sun roams
   the sky above a yucca-flowered globe
   of a rounded cliff.

A palm floating
   in the hot air
         dry like wood
   above a wood body
         immobile
   eyes clear as glass
   skin peach colored
        with sun

Below,
   far below
         children playing
            yelling
            screaming
as adults quietly, cautiously eat
   watching, worrying.

The children sing a choir of noise
    as they walk
            without work
                    out upon the water.

They are magicians
    and wave to each other
                    knowing.
The adults do not see.

15

ODE TO EDWARD ABBEY PART 1:
A DEATH IN THE FAMILY

MARCH 1989

Ed Abbey is dead.

Abbey did not and would not want to be a leader. He would have detested being worshipped, glorified and exalted. I am certain that's why he threw beer cans out car windows and drove a Cadillac. He loved to piss people off, especially those of us who dared come close to worshipping him.

But Abbey was a leader to me, at least for a while, at least until now. Abbey led me to the life I am living and loving, and his death plays a new role for me. His death dictates a new role for all of us who have found that task in life worth living for: to appreciate and to defend the Real World, wildness, wilderness.

So at the time of his death, rather than mourn him I want to celebrate his life, to share what he has done for me, and to put forth the challenge his death gives us all.

The night I heard about Abbey's death, I rounded up some friends, other Abbey-ists, and we got drunk. Under the green glow of the neon cactus at a dark local bar, we reminisced about him like he was a close friend, a family member. Even though none of us had ever met the man, he was kin. That night and in the days that followed in which I read and wrote and reflected, I was awed to discover how much of the path I am on now was guided by a map drawn by Abbey.

I had never heard of running rivers until Abbey; now I pass my summers guiding for the joy and the money. I have lived and worked in the Utah desert, my spiritual mecca (along with tens of thousands of others), in emulation of *Desert Solitaire*. I was a frustrated, aimless tree-hugger who thought he was alone in a hopeless world; through Abbey I learned of other defenders, of their anarchistic, tribal unity, and of the hope of persevering in the long run, that we will "outlive the bastards."

These paths were mapped out by Abbey, but I walk them my own way. I take my own turns and find my own terrain. Abbey would hate someone who merely mimicked and emulated him. His gift was not his role model; Abbey's gift was the gift of self: Find your self; trust your self; be true to your self and don't let anyone ever take that away from you. "Why betray common sense for the sake of any theory, cult or doctrine?" he asked.

Abbey spurred me on. "Sentiment without action is the ruin of the soul," he chided. So I dedicated my life to acting upon my sentiments. Now I find myself in graduate school, investing in a tool—learning to monkeywrench with a pen— and all directly attributable to Ed Abbey.

For myself and others he spoke to, Abbey rekindled the earth-dweller inside us, and reminded us that this is *our* planet, no matter who tells us otherwise. The power of the senses is all we need to learn this fact; our consciences are all we need to be reassured as defenders of all that is wild, healthy, alive—of the planet and therefore of ourselves—as we battle against the powerful mass delusion of industrialism, anthropocentricism, homogeny, politics, and the weakening of the human-animal through urban imprisonment and economic slavery.

> But where is home? Surely not the walled-in prison of the cities, under that low ceiling of carbon monoxide and nitrogen oxides and acid rain—the leaky malaise of an overdeveloped, overcrowded, self-destroying civilization—where most people are compelled to serve their time and please the wardens if they can. For many, for more and more of us, the out-of-doors is our true ancestral estate. For a mere five thousand years we have grubbed in the soil and laid brick upon brick to build the cities; but for a million years before that we lived the leisurely, free, and adventurous life of hunters and gatherers, warriors and tamers of horses. How can we pluck *that* deep root of feeling from the racial consciousness? Impossible.
>
> —from *Down the River*

Abbey insisted that the fight cannot be compromised by falling for a better form of servitude, but that we live and struggle for nothing less than the truest right of all people— and of animals and trees and rivers and rocks—to be free to live, to love, to explore and risk death in the joyous struggle for survival. "What's the gain," he asked, "in ridding ourselves of the Judeo-Christian hierarchy of intellectual oligarchs if we merely and meekly accept another power-hungry ideology in its place?" Or, more concisely, "Anarchy does not mean 'no rule'; it means 'no rulers'."

Abbey warned that a world of people packed together like dead fish in a can is easy to control, easy to delude. Our spirits live in wilderness, he said. Healthy, necessary diversity thrives only in unfenced, untamed, unlimited wilderness. Wilderness! We must have wilderness! And lots of it.

Why this cult of wilderness?... There are many answers, all good, each sufficient. Peace is often mentioned; beauty; spiritual refreshment, whatever that means; re-creation for the soul, whatever that is; escape; novelty, the delight of something different; truth and understanding and wisdom—commendable virtues in any man, anytime; ecology and all that, meaning the salvation of variety, diversity, possibility and potentiality, the preservation of the genetic reservoir, the answers to questions that we have not yet even learned to ask, a connection to the origin of things, an opening into the future, a source of sanity for the present—all true, all wonderful, all more than enough to answer such a dumb dead degrading question as "Why wilderness?"

To which, nevertheless, I shall append one further answer anyway: because we like the taste of freedom; because we like the smell of danger.

—from *Beyond the Wall*

The challenge handed us by Abbey's death is to pick up the fight, to nurture and spread the spirit he brought out in us, to keep his uncompromising force alive. Abbey would not want to be followed around like a general by a bunch of soldiers, no matter how well intentioned they might be, but he would want, I believe, people willing to stand next to him who are not afraid to speak and act honestly, heroically, people who are untamed and unbending. Wild people. He did not want to be a leader, but he was not afraid to be a hero. Or a fool, as long as it was honest.

The challenge is also to appreciate as well as to defend. Abbey's work is a blend of loving and fighting, anger and joy, embracing and monkey wrenching. "It is not enough to fight for the land; it is even more important to enjoy it," he reminded. We must take the time to keep a growing, healthy relationship with the earth, and with our cohabitators on this planet. Defending wilderness cannot be a knee-jerk reaction or another ideology; it must be true *defense,* a fighting from within. It must be self-defense. There would be no victory in stopping the military-industrial megamachine only to lose our true natures in the process. We must have healthy seeds to plant once we ride out this storm.

Appreciate and defend. Do not sell your self for anything. Ed Abbey is dead. For me, the one who has been my guide is gone. But I am joyous: He is off on another adventure, and I have been set free to grow and to challenge myself loving and fighting, and to apply and follow what I have been taught wherever it leads me. May I have the courage to do it as well as Abbey did.

The night I heard about Abbey's death, after my friends had gone home, I walked under the first-quarter moon. As I liberated some land from its barbed burden in Abbey's honor, I thought of a wish he made for others. Now I wish the same for him.

> May your trails be dim, lonesome, stony, narrow, winding and only slightly uphill. May the wind bring rain for the slickrock potholes fourteen miles on the other side of yonder blue ridge. May God's dog serenade your campfire, may the rattlesnake and the screech owl amuse your reveries, may the Great Sun dazzle your eyes by day and the Great Bear watch over you at night.
>
> —from *Beyond the Wall*

16

ODE TO EDWARD ABBEY PART 2: ABBEY LIVES!

Yes, we need heroes. We need heroines. But they should serve only as inspiration and examples, not as leaders.

—from "Theory of Anarchy,"
One Life at a Time, Please

MARCH 1994

Here's what the invitation says:

Announcing
the 6th Annual
Abbey Party

A celebration of life and land (what else is there?)
and the life, work and philosophy of Ed Abbey
on the fifth anniversary of his joining
the land he loves
Festivities begin at 6 p.m.
Abbey Film Festival at Midnight

I prepare for the party, which this year will be held in our garage. Other years it has been held around a big campfire, in a city apartment, around a kitchen table, and in a downtown bar. It doesn't matter much where it is. There have been as few as six attendees and as many as thirty five. That doesn't matter much, either.

I sweep the dirt and debris and oil-soaked kitty litter from the cement floor. Great dust clouds billow out into the cool air. A grey sky looms out there; rain looks likely. Or snow. I push the many boxes of junk and stacks of boating gear and loose tools and the chair I keep meaning to refinish to the sides of the two-bay garage, stacking what I can on shelves. I leave in the middle of the room the World War II-era console radio receiver, its fine finish bubbled and scarred from a house fire. It stands about three and a half feet tall with a broad top. I push it up against a beam, where it will serve as the altar.

The Abbey Altar is always the centerpiece of the Abbey Party, regardless of the party's location or attendance. The Abbey party is much like a traditional party with many of the normal party accoutrements—music and food and beer—but the Abbey Party requires some distinctive party paraphernalia, of which the altar is the most unique. I wander into the house and carry back to the garage a few special items. I lay on the altar: a quart of tequila, a potpourri of shot glasses, five uncut limes, a hunting knife for the cutting of the latter, and a salt shaker; a handful of cigars, of the "good, cheap, working man's" variety, of course; a collection of Abbey books to be available for perusal during the party, and a smaller collection of videos that will later be shown during the midnight film festival (if you're an Abbey fan, you haven't lived until you have heard Buddy Ebsen and Ron Howard rattle off classic Abbey lines); a fat candle; and behind the candle I tack to the beam a laminated picture of the mischievously grinning Abbey leaning on the butt of a shot-gun next to a blasted-out T.V.

When the first partygoers arrive the candle will be lit, the bottle of tequila cracked, and a shot poured for Abbey that will remain on the altar throughout the festivities (although it usually mysteriously disappears before dawn). Attendees gain admittance to the party by bringing a token for the altar representing what Abbey means to them, or that signifies their spirit of the outdoors, or is a talisman for their kinship with the spirit of Abbey, or...whatever. Most anything is acceptable. In past years, the morning after the party has dawned to see the altar cluttered with smudge sticks, antique monkey wrenches, old deer antlers, rotten bandanas, potted cacti, plastic carnations, poems, petitions, sprigs of sage, boughs of trees, rocks, cigar butts, photographs, hand-scrawled notes, spilled tequila, lime rinds, and so on. You get the picture.

It all seems rather silly, really. And it is. For most who come, the Abbey Party is no more than an annual excuse to gather a group of friends in the late winter or early spring, to vent some cabin-fever pressure, to drink beer and talk and build some enthusiasm for the summer-month adventures ahead. That's a lot of what the Abbey Party means to me, too. As the small print on the invitation clearly states:

(Or, put more simply, a small, friendly gathering at which traditional social niceties are dispensed with and the stated purpose is a philosophical and garrulous drunk.)

But for me, the preparing for the party is something dif-ferent. I don't mean just the sweeping of the garage floor and the setting up of the altar, I'm thinking also of the days before the Abbey Party, the time of the Abbey Party, the piece of the year, the slice of the season. This is the time that my friends, who come for the festive atmosphere and the funny altar and the midnight film festival, don't see. All year I look forward to these days, knowing that then I will pull my mental boat from the river of my life's work and pause; in this annual rite of renewal, of reflection, of introspection, I will think back on the trip so far and scout a route ahead.

Abbey is essential for this yearly work—he has always been a guide and companion on this trip for me, even if only spectrally through his writings. Every year during the time of the Abbey Party, I spend time with him again. I walk and sit and write, rearticulate my goals, reevaluate my motivations, and review my successes and failures. I touch my life thoughtfully, taking nothing for granted, which is so easy to do in the busy-ness of day-to-day business. And, importantly, I perform these rites with the spirit that I find to be common ground between myself and Abbey, a spirit of both anger and love ("How feel one without the other?" he asked). I reread some Abbey and reacquaint myself with that spirit as it is conjured up by his words. I seek answers to a few questions, pick up some advice and wisdom. Every year at this time I take a long walk with my hero.

Hero. The word sounds silly—as silly as the altar—for a grown man like me to use. It seems amusing for an adult to admit having a hero who is an eccentric and dead writer whom I never even met. But it's true. So let me explain what I mean when I say Abbey is my hero, because I don't think having heros is silly—I think it's essential.

A hero is someone who shares a similar perspective on the world, sees the same forces at work, and order values in the same way.

A hero is someone who embodies a kindred spirit, follows comfortable courses of action, bears a familiar sense of responsibility, responds to circumstances in admirable ways, draws alike conclusions from empirical reality.

A hero is these things, but is also someone further along than yourself in life in the application of conscience, perspective, and spirit; a hero has lived with them longer, applied them, tested them, learned from them, and let them evolve; and from that hero's experiences you can learn lessons.

In a nutshell, a hero is someone who is on the same compass bearing of the conscience as you are, but who has travelled further on those bearings.

Heroes are tools; their actions and reflections can be useful and important guides as we amble through our own lives. A hero is a guide, but that is all; a hero is a leader by example, not by rule. To have a hero, it is vital to remember that while you are kindred spirits, you are different people. A hero's life can't take the place of having your own experiences and challenges and tasks; to have a hero, you yourself must be living your own life. You can't blindly follow a hero's actions, motivations, or thinking; you don't imitate your hero's lifestyle, you emulate your hero's *living* style.

Why have heros? They leave crumbs on the trail, they suggest a way when you're lost, they recharge your energy when it is drained, they offer a foil with which to test possible decisions, and they can affirm your choices when you're unsure. Since they're ahead of us yet we travel in like styles, heroes carve and blaze trails for us, offer maps for life's ethical wildernesses. But still you ultimately have to go it alone, thinking, aware, reactive, alert, independent. You must follow your own conscience, develop your own perspective, you must find your own spirit.

And so it is with me and my hero, Ed Abbey.

I lean on my broom as I look at the photographed Abbey lean on the shotgun. I look at the altar, and I laugh. It is kind of silly for a guy well into his thirties to have a party for his hero. But I've been called worse things than silly. Abbey was not afraid to play the fool, as long as he was being honest to himself. Me either, I guess.

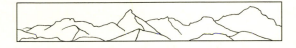

17

My Life In A nutshell

 MARVEL.

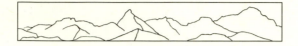

18

LET US NOW PRAISE
THE COLORADO SQUAWFISH

*T*HE COLORADO SQUAWFISH: PEOPLE EITHER HATE IT,
DON'T CARE ABOUT IT, OR USE IT TO THEIR ADVANTAGE.
Nobody loves it. With no howl, no fierce or inspiring presence, the Colorado squawfish doesn't attract the admirers or
defenders that the more glamorous Endangered Species do.

But the Colorado squawfish and its endangered kin, the
Razorback sucker, and the Bonytail and Humpback chubs,
are siblings of the wolf, grizzly, buffalo, desert tortoise, black-
footed ferret, and the spotted owl. They are all native offspring of our once-wild West.

To dam builders and water diverters, the squawfish is a
dinosaur, a "trash fish" posing an absurd obstacle to the
future. To lovers of wild rivers and opponents of expensive
water projects, the squawfish is a monkey wrench in that
river-taming machinery. But few see the Colorado squawfish
for what it IS: the embodiment of the Colorado River Basin;
the basin's unique landscape come to life.

The American West's great river system, the Colorado and
its tributaries, stands out among the world's rivers. Its basin
drains seven states and covers 246,000 square miles. In its

run to the Pacific at the Sea of Cortez, the Colorado drops more than two miles in elevation, crossing mountain, plateau, mesa and desert country. This landscape forged a river that is distinctively fast, turbulent, erosive, salty, muddy and moody.

With a demanding parent like that, the Colorado squawfish had to evolve strong and adaptive. Distinctive among the world's fish, the squawfish grew big, mobile, and mean. Squawfish weighing nearly 80 pounds and 6 feet long have been caught. One fish was radio-tracked migrating upstream almost 200 miles to spawn and find fresh habitat.

Early anglers didn't consider the squawfish a "trash fish." They called it "white salmon," "Colorado salmon," or just "salmon" because of its migratory habits—and its great taste. The Colorado Basin's native people, early settlers and first explorers all ate the abundant squawfish. Sport and commercial squawfish fisheries thrived in the early 20th century.

Today, no wild Colorado squawfish survive in the lower Colorado River. The few surviving wild squawfish live in small reaches of the upper Colorado, including in our own San Juan River, above Arizona's Glen Canyon Dam. Even there, though, the squawfish's wild river is altered; its flows are tamed, temperatures moderated, silt loads settled out and livable habitat fractured by dams, irrigation systems and diversions. And more changes loom—including the Animas-La Plata Project—that threaten to push the squawfish into history.

Forces pushing for more water development argue that water projects made the West what it is today. No doubt. Without the squawfish the West will be a multi-million-year-old landscape so beaten for the economic gain of a couple generations that it cannot support even its own offspring. But doesn't protecting this fish hurt economic development? That development landed the Colorado squawfish on the Endangered Species list in the first place; that listing signals the passing of a reasonable limit to the economic taming of the Colorado River system's wildness.

You won't see people wearing "Save the Squawfish" t-shirts or find full-page ads of cute, big-eyed squawfish fry in environmental magazines. But even if you never see this critter in your life, the Colorado squawfish and its endangered kin embody what the West was—and how at least some of it should remain.

That's why they deserve to live.

19

THOUGHTS ON
A CHILD'S FIRST RIVER TRIP

*T*HIS IS OUR THIRD DAY ON THE RIVER, AND IT IS FANTAS-
TIC. EVERY BEND IN THE CANYON IS A NEW AND
grandiose display of earthworks. This corner reveals huge
horizontal bands of white sandstone cut with half-shell
alcoves and amphitheaters; back there we were dwarfed by
river-to-canyon-top stair-stepping talus slopes. A half moon
hangs above it all like God's grin. My wife, my young son,
and I silently watch worlds unfold at the river's pace.

I lean on the oars, unnecessary for most of today's floating. The blades dry in the desert air. I just study, listen, smell.
Sarah sits on the front right tube of our cataraft. She leans
back on the seat, knees up, arms folded, face shaded but peering from a slot under her straw hat. Webb stands behind me.
He is above my head, and when I look up I see his small face
backdropped by big blue sky. He, too, stares off somewhere,
until he looks down at me looking up. His round cheeks and
blue eyes and white-blond curls comprise a serious, pensive
face. He holds my look. I smile, and he smiles back.

For young Webb, this late-spring float of Colorado's
Dolores River is his first multi-day river trip. Looking back

over these three days of floating, river-side dinners and nights filled with fat stars, I would say this venture has been blessed and inspiring and therapeutic. Certainly it has been those things for me and my wife, and for Webb I believe it has been some child's equivalent of those. But I also must admit that as splendid as this venture appears from this third-day vantage point, the decision to take Webb along was not an easy one for Sarah and me. It's not that we didn't want him here—wilderness trips and river running lie at the heart of my and Sarah's life together, and we want him to share in that life—but Webb is only a year old.

As we pondered the idea in the weeks before the trip, the risks gnawed at us like ulcers. Even on a flat and slow stretch of river such as this one, there is danger in wilderness when you're away from doctors and phones and pharmacies. Wilderness isn't child-proofed; a lot can happen to a crawling and curious child surrounded by such a new world. There is water, storms, insects and rattle snakes. There can be illness and tumbles and broken limbs. There will be campfires, cookware and river-running equipment.

And the risks weren't all we worried about; our selfish selves had something to say. Part of the reason we take wilderness trips is to break the demands of our daily routines. But how relaxing will it be for us, we wondered, to deal with our nuclear-powered one year old on a 16-foot raft for three days? As any parent can tell you, a toddler is not so much a person as it is a circumstance. People or pets you can negotiate with or at least train, but a small child is more like the weather; there's no use arguing with or trying to change what is presented. You just deal, continuously. We could picture the gory details: First he'll want to throw his toys in the river; next he'll want to crawl somewhere, anywhere, off the boat.

It didn't look good. It didn't seem smart. Our friends, once they stopped laughing and realized we were serious, urged us to stay home and rent a movie instead. Still, we decided to test our little-boy-on-the-river theory, so we took an afternoon's shakedown cruise down a languid, meandering stretch of river near our house. On a clear and dazzling Saturday afternoon, under sparkling Western skies and alongside cottonwoods and broad meadows, we floated through Hell. Webb got sunburned and frantic and cried. A lot. Sarah and I got exhausted and short-tempered and cried. A lot. Our friends got annoyed, even though they were vindicated, and haven't called us since.

We decided to go for it.

The risks still called for Alka-Seltzer. We talked a lot about those risks, and concluded that danger is a part of life, ours and Webb's, and there's no use letting them anchor us to the living room. The river, we decided, won't be much different from the other camping and road trips we had taken and survived. We would be careful and aware; we would let Webb be absorbed by and wander (as much as a one year old can) this great, big, dangerous world he has inherited, but we would watch him vigilantly. As for the parenting challenge, we opted for strapping a portable playpen to the rear cargo frame, offering the life-jacketed Webb room to play and crawl, or a place to just stand and watch the canyon roll by.

And, remarkably, that is what he has done for three days now. He has even napped, snoozing to the subtle slosh of the river in the shade of a tarp bungee-corded around the playpen's top. This is not to say that these three days haven't been work for me and Sarah; they have been. We have put in our time rowing hard upstream to retrieve his bath books and balls and yellow rubber duckies. We have passed hours entertaining him with lengthy and animated explanations of sedimentary geology and Anasazi history and other things he won't understand for years. We even let him read the river guide book for a while, which he did thoughtfully for nearly forty-five minutes. Or so we thought, until we realized he had pulled out all the relevant pages. We never found them.

It's been challenging, but we have survived. The trip has been injury-free and has been no more work, really, than chasing Webb through his normal days at home. I dare say Sarah and I are even having a great time, even though we have spent our nights at camp crawling with Webb through sand and mud and into tamarisk, while our companions in the other boats sipped cocktails and took exploratory hikes. That's okay; we can live with that small sacrifice to be out here on the Dolores River together, with our son.

Still, there is something more. Last night, as we savored some time alone together after putting Webb to bed, after we were done congratulating each other, after we thought we had overcome the little-boy-on-the-river quandaries, we were disturbed to find another stream of questions flowing to some deep questions: Who remembers anything they did when they were only a year old? I asked Sarah. Why suffer the risks and hassles when he won't even recall being here? Will our little one year old get anything out of this river trip?

I have pondered those new quandaries all day today. After rewinding and playing back these three days on the river, I conclude this little man has gotten something from this river trip. I sense he has learned and experienced from it, and even enjoyed it. This is hard to prove, of course, and the only evidence I have is his tiny laugh from the dark as Sarah and I said good-night through the tent screen, his awed squeal "*dee!*" and finger pointing in the direction of a canyon wrens' song, his two-toothed smile and hinged-finger wave to beached boats we passed, and his furrowed brow as his tiny fingers examined beach sand and splashed water while we bathed him in a warm side-creek.

Will he remember this trip? Maybe not as the "this happened and then we did that" recollections we adults conceive of as memories. Children are *feeling* critters, with soft and malleable spirits, and I suspect that over these three days something of his personality was sculpted by wind, shaped by the flow of the river, and colored by soil and bedrock.

Was it worth the risks and hassles? Today, as the river slides us along, now past a side canyon where narrow red walls bend away toward some unseen other world, I tally the costs and benefits. There were tears and bumps and meals later than he would have liked. Sarah and I passed on walks and swims and late nights around the fire we would have liked. Still, like all the experiences of raising a child, this venture was more work than expected but more rewarding than we could have hoped for. Although Webb may not talk or walk, he is a living, experiencing person, and we saw him breathe in the wilderness, the real living world. And this changed us all.

I squint up into the sun again and see Webb has returned to his toddler's study of the canyon's walls. He leans, his little face scrunched up against his little fingers that grip the crib's side. I am joyous beyond words. Perhaps bringing a child into this, immersing him in life and land, risk and hassle and all, is all a parent can do. Perhaps it is the best we can do.

20

ON HUNTING

I WAS HAVING BREAKFAST WITH A FRIEND THE OTHER DAY WHEN HE STARTED TALKING ABOUT HUNTING. While backpacking recently he had witnessed a bow hunter's deer kill, and it disturbed him deeply. Knowing my environmental sensibilities, my friend assumed he had a sympathetic ear. And he did. For any lover of wilderness and wildness, it is painful to watch any wild creature die by human hands.

But what my friend didn't know is that on the same day he viewed death, I was bow hunting.

So the question presents itself: How can an environmentalist be a hunter?

Last Saturday, as I slowly and silently walked through aspen and firs in a steady drizzle, my bow in my hand, I searched my spirit for an answer to that question. My conclusion: Hunting is one of the most wild and environmental things I do.

My reasoning does not follow the justification offered in the hunter-safety classes. There hunters are reminded about how they "pay their own way" through license fees that purchase habitat, of how we funnel money into local economies, about how hunters keep game numbers within the carrying capacity of the land.

Heaven forbid I should have to justify my hunting with dollars and cents. And I don't want to be another argument against the four-legged predators. I don't even like the term "game"; I want to be honest about what I stalk—living, breathing, wild creatures. Not game, but wildlife. And my hunting ends one of those wild lives, if I'm successful.

How can I do that? Understanding that feelings as deep as this one are impossible to do justice to in words, I still offer three answers:

Hunting is a ritual. It takes discipline, exertion and total attention. It requires skill, practice and patience. It is humbling, enlightening and educational. Stepping into the woods with a bow is a ceremony that seals the relationship between myself and these mountains because I am asking them to give me something, and I must earn it.

Hunting hones my awareness to a razor edge. Stepping deliberately, silently, painfully slowly through the woods; using each of my senses to its fullest; reading the messages deer and elk have written on the forest. Absorbing myself in these actions for hours creates an immediate-moment aliveness in me that shines for days afterward. Can't other activities do this? To some degree, but nothing brings alive our genetic animal awareness like the predator-prey dance of nature. And that dance is something I, as an environmentalist who believes we are of and not above nature, want to be part of.

We are animals, and we kill for food. Rather than hiding behind store labels and menus, I want to accept responsibility for that animal reality, honestly, with my eyes wide open. Eating beef or chicken or fish, or even wearing leather, is no less killing than venison taken with my bow. At least the deer I take lived a wild life, had a fighting chance, wasn't shot full of hormones and raised in a five-acre feed lot. At least I had a chance to thank the animal that will feed me.

Yes, I hunt. I know hunting is not for everyone. And, certainly, not all hunters are as philosophical as I am here. But I know that when I hunt I am alive, honest, and wild. When I hunt I know I don't just write about or fight for the environment, I am part of it.

21

RECLAIMING THE NEW TERRA INCOGNITA

*Y*ARMONY MOUNTAIN CALLED TO ME FOR A LONG TIME BEFORE I ANSWERED. FOR THREE SUMMERS I GUIDED rafters down the stretch of the Colorado River that winds around Yarmony. Each day, while reading the river, I also read Yarmony's sandstone face, its aspen-grove bands, its pinion and juniper shroud. I read a story of wildness.

On a clear, hot, high-desert day, I at last traded my river sandals for boots and started up the mountain. I hiked over sage flats, through forest, and up washes. As expected, I found not a trace of other humans: people who pass through this country usually confine themselves to the county road below.

Four hours later, soaked in sweat and with muscles screaming, I pulled myself over a lip and onto Yarmony's flat top. For all the effort and pain, the summit offered the grandest of rewards. I stood alone before a serrated panorama of the highest of Colorado's high country: the Gore Range, the Saguache Range, the Mount of the Holy Cross, the Flattops. I'd climbed out of the roaded river bottom up onto earth molded by sun, rain, snow, and wind, and guarded by elk, deer, mountain lions, and eagles. I'd earned entrance into this wild domain by meeting the mountain on its own terms.

Then I saw it, winding up the gentler, northern slope of Yarmony. A road. It was no interstate highway. In fact, it wasn't even paved. It was merely a lonely two-track through the sage, traveled, I'm sure, only by an occasional hunter or rancher. I knew that despite this automobile path, Yarmony was still remote, unlogged, and unmined, with its wild ecology intact.

Still, that road slapped me in the face. Just knowing it was there, realizing that anyone with wheels and an engine could buzz up and savor the same view I'd busted my butt for, turned Yarmony from a place of wild joy into just another roadside attraction.

Roads trigger an annoying Pavlovian response in me, and through many a conversation with outdoor-minded folks, I've realized I'm not alone in this affliction. We're so ingrained with the semantics of our automobile culture, that most of our conceptions and recollections of the land take on the form of a mental road map. We remember route corridors dotted with towns and sights, and we measure the lengths of these corridors in time passed at 45 or 55 or 65 miles per hour. Places outside the corridors are usually blank on our mental maps—terra firma reduced to terra incognita.

On a topographic map, however, roads become mere scratches on a more vivid and complex picture. On a topo map, it's the areas between the highways that stand out, with contour lines describing the shape, depth, height, and character of a piece of land—telling a story about a place. You can't traverse contour lines at 55 or 45 or even 10 miles an hour. You have to walk, crunching soil, grunting up hills, scaling rock, and feeling the heat of the noon sun. If an inch on the road map means an hour, it means days on the topographic map. It means wildness.

I'd approached Yarmony walking and because I thought that was the only way to the top, my brain mapped out the mountain topographically: contour lines tightening to a point, dark green flanks and a light green top, blue lines coursing down the sides. But when I saw that road, my brain fell back on my cultural training and remapped Yarmony. It flattened and turned pasty white, like the blank spot between I-70 and U.S. 40 that Yarmony occupies on my road map.

Road map thinking is squeezing whole landscapes out of our heads and altering the way we perceive places. For instance, according to my Rand McNally, here's how I got from Boulder to Yarmony mountain: I took Colorado 93 out of Boulder to U.S. 6. Stayed on U.S. 6 for half an hour to I-70. Then 70 west, through the Eisenhower Tunnel, to Colorado 9. On Route 9 for half an hour to Grand County Road 1, then drove 20 minutes on a gravel road to the bridge across the Colorado River. That brings me to the foot of Yarmony, which, of course, isn't actually on my Rand McNally.

I crossed the heart of the Rockies that day, but where in that description are the rivers, the mountains, the valleys, the sense of character that makes this land so special? Sure, I saw the natural features as I drove by, but they passed unobtrusively, like T.V. commercials—watched but not seen.

Only a perceptual revolution can stop this mental erasing of the land. Each of us can consciously turn back the pages, from the road maps of today's automobile culture to the topographic maps of a foot culture. We must make a conscious effort to sketch the lay of the land onto our brains.

Months later I returned to Yarmony. This time I drove out of Boulder following the foothills south. At Clear Creek I turned west, up Clear Creek Canyon. After crossing the Continental Divide north of Grays Peak, I descended until I reached the Blue River. I followed the Blue River downstream to the Colorado, then passed through the Gore Range to Yarmony Mountain.

I grunted to the top. The road was still there, but this time it seemed lost in an aspen stand, subdued by sagebrush. I ignored the road. In my private revolution I chose to walk, and through that effort I reclaimed Yarmony's wildness. I reclaimed the terra incognita.

22

WHY FREE-MARKET ENVIRONMENTALISM WILL FAIL

JOY AND SIMPLICITY DON'T MAKE MONEY; FUN, ADVENTURE, AND EXTREME SPORTS DO. LIVING WITHIN THE LIMITS of the land doesn't make money; structural solutions, technology, production, and population on a grand and ever-grander scale do.

23

A MARRIED MAN

Got a woman I love and she loves me
And we live on a piece of land
I never know quite how to measure these things
But I guess I'm a happy man.

—Bruce Cockburn
"Great Big Love"

*T*OMORROW, TOM LEAVES FOR MEXICO. SO TONIGHT WE SHARE A COLD TWELVE PACK UNDER A COLD JANUARY sky. With the temperature in the low 20s, we sit like a couple of watchmen on the bridge of an icebreaker, wrapped in thick jackets, gloves, and hats From a beat-up sofa on a friend's porch—our friend isn't home—we keep watch over a dark side street in Durango, Colorado.

Tomorrow, Tom will get a ride down through New Mexico to Nogales. There he will hop a train to Mexico City, and begin a four-month hitchhiking journey through Latin America.

Most people faced with imminent departure to another country have their minds full, double-checking plans, scanning mental lists. Did I remember the fluoride toothpaste? Underwear? The Swiss Army knife? Instead, Tom sits here under his tall, grey wool cap, hands clutched together around his beer can to keep it from freezing, his boots hiked up on the porch rail, and he talks about my situation.

"I can't believe you're going to have a kid," he says. He follows this with a deep swig of beer, like he's rinsing those funky tasting words from his mouth.

Tom is haunted. He wears the same leery, troubled expression Donald Sutherland did in *Invasion of the Body Snatchers*. As Tom looks at me, he seems to be thinking that I look like his friend, the one who, on a whim, might have thrown some essentials into a backpack and joined his southern migration for the winter; that I look like the traveling companion who shared his love for walking across little-appreciated corners of the map; that I look the same, with the K-Mart™ work boots, Red Sox cap, and beard that never seems quite finished. But to Tom, it seems like his friend's body has been snatched.

Who inhabits his friend's body? Some alien who is married, owns a house, and now is going to sprout an offspring. A stranger who traded in his walking shoes for ankle irons, and who seems to be enjoying it.

"Do you know what you're doing?" he has asked me a few times. Sometimes the question is posed silently, just bunched eyebrows when I rattle off some of my life plans. Other times he asks outright, like tonight.

"Do you know what you're doing?"

I don't answer. Together we silently look out across the street at a yellow and white-trimmed Victorian house, a survivor from Durango's mining heyday.

"I can't imagine having a kid. Or a wife," he says, finally breaking the silence. His frosty breath dissipates, but the words hang in the air.

I respond with a sip of beer and a glance out over the trenches that January's snow has made of the streets. A big, old luxury car whose heyday has also passed rolls by, crunching crusty snow. It turns left after it passes us, and silence returns.

Tom may worry about me, but I don't worry about him. I don't fret about his safety, his ability to navigate in a foreign land, or his social skills. Well, maybe a little about his social skills, as he slides into the dark space next to the house to relieve himself. But I'm not concerned that he won't eat right or find good water. I don't worry because my friend here is a traveler, and I have faith that his traveler's spirit will guide him true.

And he shouldn't worry about me, because I am—still—a traveler, too.

Sure, these days I don't cover as much ground as he does. The only time I went to Mexico I swam, across the Rio Grande. I stood there among the agave and Mexican cowpies for a few momentitos, then swam back. Back here in Colorado, Tom has scaled and viewed the scene from hundreds of peaks more than I have. And I don't hitch-hike much since the afternoon I waited quietly in the back seat of a Delta 88, my feet resting on beer cans and a shot gun, while two drunken madmen debated whether to drop me off or not. I think that was when my walking career really took off.

Tom lives light, owns little, and is always headed out, out there, somewhere. I run into him at City Market sometimes, where he's stuffing a loaf of bread, a quart of milk, a box of tea, and a jar of peanut butter into his scarred and patched backpack, the same pack he carried up Denali. He dresses the same every day—Timberland hikers, red Gore-Tex parka, big-brimmed leather hat—whether he's walking the to the bakery for coffee or along the Divide toward the Indian Peaks. And many Friday afternoons, he's standing in front of the Conoco on the edge of town, thumb out trolling for a ride to Arizona or Mexico or Utah.

Me? Well, I stick around.

Every morning I walk down my dirt driveway to the pavement, which I follow around the corner into the valley where the elk winter. I check the flow in the ditch, I inspect the foliage on the scrub oaks, I examine the color of the sandstone bluffs. In the afternoon I often return there, sauntering along the same route. On weekends, I usually make it up some nearby hill, like the rampart of cliffs behind our house, or the unnamed ridge ribboned with logging roads on the other side of the valley. Sometimes my wife, Sarah, and I get into the truck and venture into the high country, where I pull out the fishing pole for a few hours or where we walk up above treeline and across the tundra. And if we can string a few days together, then we stuff our packs with sleeping bags and trail mix and venture into the wilderness to pass a few nights under the Milky Way.

But I rarely get too far from home anymore. I don't want to. Tom doesn't understand this.

But I think I do: Tom and I may now be walking different trails in life, but we are still both walkers. Kindred spirits, we are. Even though he is headed for Mexico while I'm headed to County Road 205, we are both travelers; we both want to learn, to experience, to see what others miss.

The difference between us? I phrase it this way: Tom is a bachelor traveler, and I am a married traveler.

I don't throw out that phrase as a clever literary device; it's more than a mere metaphor or analogy. Tom flirts with lots of places, and is always courting new ones; he is a philanderer, with the emphasis on land. But me, I have made a commitment to a certain landscape. I have offered the same vows of loyalty, fidelity, and obligation to the place I live that I made when I slid a gold ring on my flesh-and-blood wife's finger. *I will stay with you, I will learn about you, I will accept you for who you are, I will stand and defend you.*

I am a married man: I am wedded to the Colorado Plateau, and to its mountains—the San Juans, the La Platas, the Abajos, and the Sleeping Ute—to the rivers—the

Animas, the San Juan, the Dolores, the Piedra—and to the canyon country weaving it all together. I am also bound to the landscape I encounter daily, the streets, hills, mesas, and foothills that surround my town. I am connected to the people who live, play, work, struggle and muddle through here, through the hot, dry summers and the snowy, cold winters, faithfully, like good spouses should. And like any marriage, my terrestrial relationship is ever-evolving, sometimes moody, and often routine. It's work. Every day I must rally the energy to seek the new and relish the familiar.

But I don't say any of this to this guy next to me. And he is quiet as he hands me another beer. But through the silence I am speared by a telepathic barb: Hey, cool marriage thing, hombre, but are you still a traveler?

I pull my cap lower over my head, to keep my thoughts in and the deepening cold out. Next to the house, a slough of snow falls from the naked branches of an elm tree, a branchalanche.

Yes, my friend, yes, my mind answers his mental query. I travel every day. I have only changed the direction of my travel. When I married Sarah, I traded a breadth of relationships for a depth of relationship; so, too, I now explore my chosen terrain deeply, rather than covering a lot of terrain. And I walk these deepening trails with a traveler's spirit: I hunger for awareness, adventure, knowledge, and challenge.

I have not been disappointed.

Should I say these things to Tom? Ramble on to him about my married-to-a-big-ol'-plateau philosophy? I think he might understand. I think he might even try it, if that were all there was to it. What really hangs him up, though, what makes my compadre here grab his pack and mumble "adios!" is the traditional marriage-to-a-person that seems to accompany squatting in one place too long; it's the little bambino on the way that kicks him down the road.

How could I explain to him that my wife and our fledgling family are an inseparable part of my marriage to this place?

I could be technical: "Place," I might say, "is the land, but it is also the community and the people that live there..."

No. I don't say that.

Philosophy might work: "Love is a skill," I would begin. "It's not a thing, sitting there, like a pond or patch of poison ivy that you just 'fall into'..."

But that wouldn't work either.

I take a sip of beer and swirl it around in my mouth, savoring the bitterness. Little ice chunks shock my teeth.

Maybe I could tell a story: "My father loved to go to the woods of northern New England," I would reminisce, then I would tell him about how up there in the hardwoods, on ancient, rounded mountains, around lakes and creeks, he taught me to fish and hunt and walk. We walked everywhere up there. While we carried our bows and fishing rods, sought out deer and brook trout, we walked. While we walked, my father told me about the land around us, about how slopes with bedrock outcroppings meant good trout pools, about how an autumn-yellow beech grove meant white-tail scrounging for beech nuts.

As we approached those places, my father grew respectful, pensive, and alert like I never saw him anywhere else. He slowed his pace, moved precisely, stepped deliberately. I imitated him as he slid each leg forward, touched his toes to the ground, and rolled his foot flat so the leaves made no sound. It was then I learned how to walk. It was then I realized that I had, all along, on all our ventures, learned all I knew about the land by how I passed through it.

I want to show that to my child.

Tom, who leaves for Mexico in the morning, hands me another beer. I just smile, nod, and don't say anything.

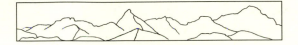

24

Time To Get Out

*M*Y LINE OF WORK DEMANDS I READ A LOT OF ENVI-
RONMENTAL JOURNALISM AND NATURE WRITING,
that I follow a number of local and regional land-use issues, that
I keep in contact with activists and players on a variety of fronts.

After a while, it gets pretty depressing.

That's the nature of the beast. Let's face it, even if you just
read the local paper, even here, in relatively remote Durango,
Colorado, you get so bombarded with news and points of view
and facts and figures about growth, tourism, subdivisions, water
projects, new golf courses and condominiums, debates over land
use and resource battles, that after a while the walls close in.

You sit at a traffic light in town and it looks like people
everywhere—they're all you can see—and the news keeps
reminding you that it's getting worse. You can't help but feel
that all is lost, that everything is used or used up, that people
and their bickering are everywhere, and they're all moving
here, wherever you are.

That's when it's time to get out.

So that's what I did. On a recent Thursday morning, two
friends and I tossed fat backpacks into the back of my truck
and we headed north, up out of the valley, to the mountains.

We outlined a route that connected some blank spots on our mental maps, and we walked for days. Just to see what's there. Just to see what's left.

And you know what? There's a lot left.

I remember now a smorgasbord of images:

Three of us sit around a small but dancing campfire. We smoke big cigars, Greg tells fish stories, and we watch the stars arrange themselves into familiar figures.

Sunrise over a serrated line of blue peaks. I sit up in my sleeping bag and watch, and do nothing else.

I look up from the trail and am startled by stark colors in vertical relief: white snow, red rock, green tundra, grey caprock, blue sky.

It's the third day out and we see the first other people since we left. They smile and wave.

We walk through a thick forest in a fantastic valley. "Wouldn't it suck if there were a road here?" Tim observes. I don't bother replying.

Elk prints over Tim's boot prints. I realize this elk must have walked the trail only minutes ahead of me. Then I look up, only ten feet from a big elk calf who looks at me like... well, like how I might look at an R.V. with a 4X4 in tow that just cut me off.

Standing on a pass under an unblemished high-country sky, I absorb 360 degrees of stunning openness—the La Platas to Missionary Ridge to the Needles to the Weminuche to Engineer to Grizzly to the San Miguel and Rico ranges. It's almost too grand to take in, too grandiose to believe. I could spend my whole life exploring just the area I can see from this perch. Maybe I will.

We, of course, had to come home. But now when the news and arguments and lawsuits and staggering statistics start to hammer my spirit into a fetal position, I draw on these memories, and they recharge me. From the valley floor I catch glimpses of peaks I have visited and they give me perspective. They tell me all is not lost. They remind me I can get out.

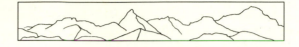

25

IN THE FOREST

I COULD LIVE HERE. THE NORTH COUNTRY OF NORTH-
ERN MINNESOTA, WISCONSIN, AND UPPER PENINSULA
Michigan has for me an allure that I am hard-pressed to
explain. I have an affinity for its dampness, frequent rains,
abundance of standing water, thick and diverse forests, and
flat topography. I believe we could be friends. Lovers. Life-
long companions.

But I am married to another place. So with the North
Country I settle for a long-distance relationship and an
annual fling. Once a year or so, my wife (the human one) and
I (and now our little boy, too) pack up whatever vehicle we are
driving that year and head east and north. Across the Rockies,
onto the Great Plains, and up into the forest. After we get
north of Minneapolis and across the Saint Croix River, the
viewing distances shorten as the trees close in, and the farms
and towns get smaller and farther apart as bogs and swamps
and lakes appear more often. The land and the little communities begin to reek of rural character—small, simple, lived in.

"That's what I like about North Country towns," Sarah
mused once as we drove through another little cluster of
buildings, "four lakes, twenty bars, and five houses."

And a bait shop, I would add, since fishing is religion up here. Friday night fish fries are the weekly service, and the bread is broken over a walleye dinner with the local brew. Up North, the people are big, the winters are long, and the communities are built around the land, whether the focus be logging, fishing, snowmobiling, or outfitting. I like it here.

This summer, though, the primary target of our venture is not the culture of the North. This year we're leaving the towns and the fish fries and are headed to a special place. After a few days visiting with some family and friends, after dropping off little Webb with Grandma and Grandpa, Sarah and I are off, alone.

✦ ✦ ✦ ✦

The first evening finds us sitting on the polished surface of a calm lake, listening to a series of reverberating calls. They greet us, welcome us to the lake, or so my ego thinks. The call comes again in its distinctive quivering tones. It varies, but this time I think it sounds like a whale imitating a wolf. Whatever this song sounds like, it is ultimately unique; the call of the loon is as much the voice of the North Country as the cry of the coyote is the song of the West.

I scan the darkening lake with my binoculars and locate a flotilla of three loons out there, black periscope heads on speckled bodies trailing wakes like silver strings. I set the binoculars down and resume the waiting. Silently, a short distance outside the cover of the lakeside maples, we anticipate the distant hootings.

As the evening's mosquitoes come to dine we grow restless, so we head farther out onto the lake, paddling our canoe as noiselessly as we can. Loons continue to call every minute or two from unpredictable directions. I have never heard so many. We clear a rocky point and move into a little bay, its center silver as the sky and its outer edge black as the woods. We glide for a few minutes, Sarah resting her paddle across her knees. I pick up my rod and cast toward shore with a splashing, popping surface lure. This occupies my hands, but my eyes devour the place.

We are in wilderness—legally designated Wilderness—yet I can see a sign of the outside world: above the far shore a flashing, red-eyed radio tower pokes up over the trees. But this doesn't bother me; it means that this Wilderness, this set-aside piece of the wild North Country, co-exists with a very lived-in landscape. This is good. As writer and farmer Wendell Berry says wilderness should feel like, and should be, just an extension of our day-to-day lives and environment. In a psychologically and physically healthy world, wilderness wouldn't be some freak thing *out there*. We need more of wildernesses alongside radio towers; we need a web of wildernesses within our web of highways and electronic networks and communities. In New England, where I was born and raised, they need about 1,500 of these places.

But they are few and far between east of the Rockies. This place we paddle through this evening is a rarity, a gem, one of the last of its kind. Having grown up in New England, I never thought I would stand in an ancient Northern forest. The landscape there had been stripped of nearly all its forest cover more than a hundred years before I was born. The ship-building industry, the rise of factories and a booming population all needed wood, and the Northern timber empires met that demand thoroughly and relentlessly.

New England's wet climate and fertile soils have allowed the forests there to stage a remarkable comeback, and those new forests are the forests I was weaned on. But they are second and third-growth stands, still maturing, still working their way through the stages of forest succession. They are beautiful, they are alive, and I love them, but they are not the ancient forests of New England. I just assumed I would never see the great maple groves, the giant yellow birches, the sky-scraping white pines, the fat and shady hemlocks that ruled the land from the retreat of the glaciers to the arrival of the cross-cut saw.

Then I heard about Sylvania. In the Upper Peninsula of Michigan—another place where the giant timber companies laid waste to the Northern forests—a piece of that ancient landscape remains. Here stands 30 square miles of never-logged forest and a series of pristine lakes so pure you can drink from them and so clear you can see boulders and sunken timber more than 50 feet down.

This tiny gem remains because a few wealthy timber barons wanted their own private place to hunt and fish in the fantastic forests they were getting rich from destroying, and because in the 1960s the Sylvania Club's grounds were purchased by the U.S. Forest Service. This acquisition has proven timely. The North Country of Wisconsin and Michigan, like rural New England, like the interior West, like my home of southwestern Colorado, is suffering a tidal wave of urban exiles seeking the small-town life and second-home getaways. Sylvania Wilderness surely would be Sylvania Estates today if it weren't already owned by all Americans as public land. And that's why Sarah and I can sit here tonight on this still lake and listen to loon calls.

I reel in my line and lay the pole in the canoe. The mosquitoes have tracked us down, so we return to our paddles. With deliberate and silent strokes we move our rented boat across the glassy and darkening surface flanked by the black curtain of shore. Bats play chicken with our heads. Four more loons glide by, dark shapes on the Confederate-gray lake. In this last breath of dusk they seem more like promises than reality—promises of something wild and new in the days and nights ahead for us here in Sylvania.

✦ ✦ ✦ ✦

The next morning we are eager to cash in on that promise. After tea and cold cereal we head out to explore some of the dozens of lakes in Sylvania. Most of those lakes are accessible to canoes over mostly short portages. Most, that is, except for today's first two portages—from Whitefish Lake, where we're camped, to tiny Hay Lake, a fine-looking bass lake the color of cranberry tea and lined with lily pads and deadfalls; then from Hay Lake to Glimmerglass Lake. These two portages are the longest in the entire Wilderness and cross the most difficult trails. On these first portages of this five-day trip, I experiment with different methods of carrying the canoe on my shoulders with the portaging yoke. This is new labor for me; Sarah and I are river rafters and kayakers, canoeing rarely and never portaging. By the time I get to Glimmerglass, I can see why. Or feel why, deep in my shoulders. This is a sick sport. This is why God invented roof racks and Jeep Wagoneers.

The next lake crossing, between Glimmerglass and Clark Lake, is easier: a shallow channel that requires only wading and pulling the canoe. Already I find myself too happy there is no portage, an ominous sign.

Alas, this whining is why we go into wilderness occasionally, as often as we can but not often enough. We go into wilderness to remind ourselves we are getting lazy, are removed from the real effort life demands, that living is work and that the work of living is not easy but is still good. We acquire from wilderness ventures spiritual tools that are useful in all our living: senses of mystery, awe, fascination, curiosity and humbleness; personal fortitude and patience; awareness of minute detail; respect for our surroundings; a presence in the immediate awareness, a sensual absorption of the immediate moment that provides the proper perspective to daily life's petty annoyances, trivial tasks, and endless thinking thinking thinking.

It must be the fortitude and patience that putting a heavy aluminum canoe on your shoulders for a walk in the woods offers. The other wilderness gifts lie in what happens next. We paddle across the south end of Clark Lake, shadowing the green ring of summer maples, moving past a small island. Off the northeast point of the island we have a close encounter with a loon.

While most water birds hit the surface from the air, this loon arrives like a surfacing submarine. Loons can fly, of course, but they can also submerge, flying underwater to fish and dodge danger for as long as five minutes and to depths of 200 feet. A peculiar survival strategy for a bird to evolve, but it must work, for the loon is the world's oldest living bird species, hanging around lake country for more than 20 million years. This age makes them unique: They are solid boned and heavy, and their legs are set far back on their bodies. These factors, improved in later models of birds, make them great swimmers but basically worthless on land, where they have to push themselves along slowly and awkwardly, sliding on their breasts. Their weight also makes them awkward in the air (most bird species evolved lightweight hollow bones). Carrying all that solid-bone weight, loons need long runs of lake surface, up to a hundred yards, to get airborne.

Unfortunately, a loon flapping and splashing and running its fat little legs off across the surface of a lake to achieve lift-off has a hard time outrunning shot-gun fire. Being evolutionary Edsels hurts them in other ways, too. Since loons nest on land, their poor land mobility makes them vulnerable to prey. They compensate by nesting in remote areas close to the water, islands and lake shores, and in the winter on coastal waterways. But this in turn makes them vulnerable to new predators that also love this habitat: second-home and resort development. And so it is that after 20 million years, only in the 20th century has extinction become a possibility for loons, as birdshot, habitat loss, pollution, and development of their coastal wintering grounds hammer them in ways evolution could not have prepared them for.

As their numbers dwindle, their mystique grows. (If they get close enough to extinction they may even get a professional sports team named after them.) I grew up on the shore lake in central New England, where loons are rare. I can remember only a couple distinct occasions when nightfall would be accompanied by that ghostly yodel from somewhere out on the black lake. My parents would come into my room and wake me, make me listen, tell me how rare it was. I would lie there after they left and continue listening. I didn't know what a loon looked like, so the sound for me became the voice of the lake I knew and loved. I felt it was a celebration of and by the lake itself that I just caught a whisper of, but would never understand. As a kid, that call was for me the soundtrack of mystery. It still is.

This loon is probably fishing, because it obviously doesn't think of us as a danger. It bobs in profile ten feet from us as we stare, admiring this view of a traditionally stand-offish critter. We note the details: the glowing red eye in a black-velvet cowling; the vertical-striped collar; the black-and-white-checked body, like a tablecloth draped over its flat back. Suddenly, the loon lifts and shakes its wings dry with a few sharp snaps; like a kayaker doing an endo, it flashes its white belly while keeping its tail feathers underwater. Then it settles again, and without a glance at us slips underwater.

It leaves us exhilarated, and we consider our encounter the appropriate preamble to our next destination, Loon Lake. We paddle over to the southeast shore, where a big, birch-lined clearing marks where the Sylvania Club's lodges once stood. One of these log cabins was 8,000 square feet and had indoor tennis courts. These were Sylvania's first and only trophy homes, like the gaudy and inflated country palaces rising across rural America of the 1990s. After the Forest Service acquired the Sylvania Club, the agency burned these buildings so the wilderness could recover, a truly worthwhile government occupation I'd happily give my tax dollars to support.

Alas, I see the Forest Service's torching work here as a noble but token effort. These lakes are saved but the rest of rural America is under siege. How often any more do you see a beautiful maple and birch-lined lake with no houses on it?

In our own Rocky Mountain West, how often do you see a lowland valley without a realtor's sign announcing an imminent subdivision? Rarely, and it's getting rarer. The Rural Revival is on as urban and suburban refugees flee their messed nests and seek the good life they see others partaking in. This is America, so if you can afford it you can have it; but those who can afford it are making it too expensive for those who had it. The countryside heads to market and grows more valuable and expensive, and the places left to live simply, cheaply, and with abundant landscape shrink. Those who want the joy of a rural lifestyle are driven out by those who just want the *style* without the effort of the *life*.

That's the disease. The symptoms are the trophy homes— dwellings whose size and style and location reflect not a need for shelter or function or appreciation of place, but the owners' desire to display their opulence—and "country style" subdivisions that are rising on the land like boils on a leper. As if more lessons were needed, these trophy homes and subdivisions confirm the fact that their money can't buy the owners a healthy, personally fulfilling philosophy, only a location where they can be seen living where the philosophy they seek once flourished. These people sacrifice nothing (but a fine piece of land); they have done no real work. They live on a place, not in it.

Stop thinking this stuff! I order myself. Appreciate where you are, man. Focus on work. We unload and portage a neck of woods to Loon Lake. As rare as undeveloped lakes are, Loon Lake is the fifth I've paddled on today in this special place. Good stuff. Good enough for a lunch spot, and I take my lunch spots seriously. On the stony and driftwood-lined shore we face south, into a cool breeze. We nibble sausage and cheese on tortillas and drink lemonade from water bottles. We look out over a rippled lake surface the same shade of grey as the sky, the two separated by a dark ragged tear, a band of trees marking the horizon. Lake and forest and sky—

that is the whole world here. And here they blend together, feeding off each other, one bleeding into the other. The sky bleeds color into the lakes, the forest bleeds trees into the lakes. The trees live in the forest and then die, falling across the forest floor to feed the forest itself, or falling from the forest across the shore and bleaching in the sun, or falling into the lake to nurse the lake life. We see dead trees around our lunch spot in all these positions. When we paddle and lean over, peering into the clear water, we can see deadfalls down there fifty feet or more. Lots of death, a sign of health.

Lunch done, we paddle across Loon Lake to another portage that will take us (actually the portages don't take us anywhere, we do all the work) to Deer Island Lake, where the big bass are supposed to hang. It's a fairly long portage, and it's hard. How long is it? I'm not sure. The distance is marked on my map, but it is measured in "rods." What the hell is a rod? I have no idea, but about halfway through the portage it occurs to me that portages are measured in rods because you really don't want to know how far they are. This one is far enough that it makes me grunt and groan and piss and moan, which I don't want to do out here. Three quarters of the way through I make an executive decision: Next time we get a smaller canoe. The outfitter gave us a big, 17-foot aluminum whale, like the ones the Navy uses to sight in howitzers. It's massive. I take a break, and to take my mind off the scream in my shoulders I read just how massive it is: The label stamped on one end says this canoe can hold six people at 150 pounds each, or 975 pounds. If you exceed that capacity, the mass of the canoe collapses in on itself, forming a black hole.

These portages, as unhappy as they make me, do serve a vital function for the land. Loon and Deer Island lakes, and several of the other lakes in this part of the park, are less visited than Clark and Crooked lakes, but are busier than Whitefish Lake. These variations are attributable to the portages. Clark and Crooked lakes are accessible to cars; to get to Whitefish Lake from your car you have to carry your canoe three-quarters of a mile, or you can cross from Clark Lake across one of the longest portages in the park. Like everywhere, work and difficulty are natural barriers to growth; combine that with an agreement to not allow machines that would obliterate the work and difficulty, and you have preservation and people. This is a law of nature, and should be a law of the nation.

Thanks to this law, on Deer Island Lake we see only two other canoes, one with a father and his two teen-age sons (who carried the canoe, I'm sure), and a lone fisherman with an ultra-light solo canoe (he gloatingly lets me lift it with one hand, all of thirty pounds, the bastard). We paddle over to the tip of Deer Island, a couple-dozen-acre hill rising from the middle of the lake and covered by a thick mixed-stand of trees. Could there really be deer out here? Perhaps, but I see no sign from the canoe, and I have no spare energy to check it out. So we relax. Actually, we loaf, a more extended, practiced, intentional art than mere relaxing. Loafing is deliberately silencing the thoughts and filling the vacuum formed with only the senses. Loafing is good, healthy, and far too unpracticed in our culture. We are afraid of being loafers, paranoid of being charged with laziness.

Not me. Loafing is what I do. It is the primary, guiding, inclusive, binding activity in my life. I like to just sit and look, and it is usually what I am really doing no matter what it may look like I am doing. I listen, and smell, and feel the minor plays of air currents on my skin. This is why I prefer float-boating to whitewater running, walking to mountain biking, telemarking and cross-country to alpine skiing, sitting on a porch sipping a cup of tea to sitting in a bar with a rock band, etc.

So we float and loaf and admire our good fortune to be here, on this lake, on this fine grey day. As I have said before, this is a rare treat for us. In our own Four Corners area if we want a "lake" experience we must make do with either piddly stock ponds or obscene Bureau of Reclamation reservoirs, which the engineers egotistically call "lakes." Only God can make a lake; our fake desert water holes are feeble replacements for a natural body of standing water in the North Country. Lakes are good, but most reservoirs suck—another law of nature.

I make the effort to peer over the gunnel. The water is so clear and deep it is black, like space. I lift my rod and take a few casts, and soon catch a nice-sized smallmouth bass. Sarah takes a picture and I let the bass go. I have made my point. We return to our loafing.

I let my senses wash over me. Two ravens squawk overhead (a familiar sound), the lake-top air is thick and creamy, scattered bird song bursts from 3-D space, cloud patterns overhead paint vibrating patterns on the water, fingers of breeze stroke my cheek. And then there's the feeling.... I can talk about what I see and do, but the feelings that emanate from this place lie beyond the rational, are too ethereal for a net of material words. Poetry may best capture what I want to say, but I am no poet. But someone else's poem comes to mind, "Northern Lake," by poet Gregory Moore:

> A mystery, from the muck
> and blooming ooze of the bottom
> to the indistinct mirror
>
> reflecting uncertain skies...
> comfort and source only to insects
> adrift amidst lily pad islands.
> Refuge of loon and muskellunge.
>
> I know only this:
> a beckoning beyond the magnetic
> pull of Polaris on summer nights...
> still more than the grip
> of ancestry, like current

on the paddle of a canoe...

a whisper, from the northland
to be lost in birch-forests,
returning only in the waves
softly received onshore
from the cut-glass landings
of teal on an unnamed lake.

Yeah. That's it.

Time to move. We lift our paddles, turn the bow to circle Deer Island, and take the first long strokes that will
return us to camp.

✦ ✦ ✦ ✦

Night: Rain threatens like the bad attitude of a steely sky,
but we sit outside and try to light a fire. Try and try again. I'm
spoiled by the West, especially in our recent drought years.
Here, even though it hasn't rained in three days, even the dri-
est leaves and twigs I can find refuse to hold a flame. I must
remember the old Eastern tricks and take the time to do it
right. I pull paper-bark from deadfall birches, I break kindling
from driftwood trees, I build an intricate carbureting pyramid
that balances fuel and air, and soon we get a small blaze going.
Next to this we stack the rest of our fuel wood to dry.

On the stove I boil two cups of hot cocoa, then stash the
mix with the rest of the food in the bear bag, which I then
hang from a distant tree. No need to taunt the locals. Then
we sit by our little fire and we study it, savor our cocoa, talk
some. We do this until the cocoa is gone, the fire just weak
coals, and we are at last tired enough to hold hands and walk
to the tent.

✦ ✦ ✦ ✦

The next morning is still grey, but there is nothing better
to brighten it than Sarah grilling up pancakes thick and rich
and sweet with wild raspberries. These berries are everywhere
in this bumper-crop year. They hang off the bushes all

around camp, all around Sylvania, like overdressed Christmas trees. We can't walk the 25 yards between the camp and the shore without filling our mouths with them. They're delicious, and they make the pancakes something spectacular.

And they fuel us for a walk. Breakfast over, food stowed and rehung, we pack our day packs in the grey and wind of the day. The wind is fragrant. It smells organic, like soil, like gaseous humus. We expect rain, and so pull on our rain gear—you can enjoy any weather if you're prepared and accept it. Since when did warm, sunny days become the only weather people can be happy with? *Y'know, I can't stand those geeky T.V. weather forecasters always apologizing and whining about rainy days hot days cloudy days windy days, sheeit...!* I preach to Sarah as I stuff dry socks into my pack. She smiles and pulls her arms through her pack straps. Once again I thank the gods for finding me someone who would put up with me. I wonder if she feels the same.

A steady drizzle begins to fall as we walk the shoreline birches to the portage, an old two-track road left over from the Sylvania Club days. Three decades later it is well on its way to being reclaimed by grasses, maples, and mosses. It takes us into the forest.

In the rainy grey light the forest is vertical lines of ashen limbs and trunks painted on a lime-green canvas. It's nice to be able to take in the whole scene from under a cap visor rather than staring at only the forest floor from the inside of an overturned canoe. I stop and look around: to my left, two adolescent hemlocks; to my right, a group of equally young (maybe sixty feet tall) maples; in front of me, a grand and old yellow birch, fat and straight and reaching up and out of sight into the leafy canopy. Its trunk is gnarled and scarred, lending it character, an air of experienced wisdom. Or so it seems to me, an over-thinking human standing humbly at its feet.

How to describe this forest we walk through? Once again, I find an adequate vocabulary elusive, like trying to throw ropes around the day. This woods is abundance and variety beyond words. I could turn it into a taxonomic list, a book-length chain of details, an extended enumeration of species frosted with a heap of adjectives, but that would still miss what I sense walking along within this world, it would still miss some knowledge that is directly given without thinking, that emanates from the ground. The point is that this is the most amazing forest I have ever seen, ever stood within, ever passed through, in my whole life.

Incredible.

All 30 square miles of it.

And here come those thoughts again.... My God, what have we done? What have we done to make 30 square miles of water, humus, and trees—together a *forest*—such a rarity, such a once-in-a-lifetime experience even for someone who has spent his life wandering forests?

Think about it: *What a continent this must have been.*

What have we done? More importantly, what are we doing? When you see what is here, when you get a feel for how little of this—of what was—is left, we must ask ourselves, how can we let the destruction continue now, today, in the Pacific Northwest, in British Columbia, in the few remaining roadless areas in the Rocky Mountains?

It is tragic and frightening what was done to New England and to the North Woods of Wisconsin, Michigan, and Minnesota, but we seem to spend all our time mourning and complaining and writing this destruction off to forces outside our control. We are content to sit on our asses and blame machines and corporations, the corporate leaders and politicians. We blame "the inevitable" and "human nature" and "progress" for devouring the landscape we claim to love, that we see so little of, that our children's children may never see.

That's bullshit. We are a nation of cowards. We are happy to be slaves when we could stand up for something; we are satisfied to have toys when we could have the world. Thoreau said, "Let us be men first, and subjects afterward," but we are happy to be subjects because it's easy and that way we don't risk losing the bones we're tossed (our cars, our T.V.s, our c.d.s, our weekends, our two-weeks-each-year vacations).

The worst part is we won't even admit our cowardice and subjugation. As individuals, we will sacrifice little or nothing—certainly not our jobs or lifestyles, and rarely even our T.V. time or beer money—to stand up for something. That is enslavement. We enslave ourselves. The machines and corporations and the growth-driven economy operate as they do because we let them through our refusal to speak up, take action, take control of them and make them back off. Machines and money—and corporations, the cross-breeding of the two—are our tools, but instead they have turned us into the tools.

I am not optimistic that we as a people will change. But I hold out a pitifully puny hope that individuals can change, can restore perspective. All change requires is action. You must act. You can't do everything (our favorite alibi), but you still must do *something*. There are a variety of strategies and roles to fill in the struggle for a healthy planet, an ethical personal life, and a just society.

For example, my friends Julie and Greg strive for a spot off the energy grid, wanting to contribute as little as possible to the carnivorous consumption of resources the grid demands. They have sacrificed a lot of the comfort and securities we in this society take for granted and instead have chosen to live in a tepee, a tiny hand-built cabin, and are now building an energy-efficient home from recycled materials. In return they not only get a sense of personal independence and freedom from industrial and economic society's tentacles, but also the comforting knowledge that they have accomplished something tangible and valuable.

Another example: I know a guy who has dedicated his life to living light materially and financially. With this lightness,

Brett walks, and his walking is how he caresses the land he loves. He physically canvasses the great landscape many of us only wish we could see more of, that we dream about, that we might get out to see on weekends or, if we're lucky, a week or two a year. He has constructed his life so he can see it, touch it, breath it all, frequently, and for great lengths of time. His reward is a depth of knowledge, a richness of understanding, and a relationship with the land the rest of us can only imagine.

Another friend of mine is a warrior. Michael focuses his skills, energy, and anger toward a mission: He wants the exterminated gray wolf reintroduced to Colorado. This is his life's work, and to do that he has constructed his life strategically toward that goal, as a warrior must. He lives a lean and mobile life; he is relentless and absorbed; he battles in the worlds of science and politics and the media. He is dedicated to a worthy and valuable, focused and defined task, and in return he receives the abandon and satisfaction that a life dedicated to a cause offers. Michael carries the aura of a warrior, a person who knows what he must do, believes it is a Good Thing, and does it.

These people are heroes in their own realms, and I admire and respect them more than they will ever know. They are doing something. But we can't all do what they do. I, myself, am not a warrior or a loner, and I am plugged into the grid; I work for change from within our society. I am drawn to writing, to family, to community, and to finding a place and making a stand on it. I may be apologizing, copping out for not sacrificing more, merely making excuses for what I am too scared or lazy to do—I have been accused of that—but I don't believe so. We must do what we do best, do it as best we can, and support others who do the same but in their individual ways. But we must do something. If enough people just do what they can within their spheres of influence, even if every one even just claimed a piece of land and defended it, with enough individuals we could be a lever against the bulk of the machine.

This is pretty hopeful stuff. And even if there is a sudden and massive awakening, a wave of people acting and standing up, it's going to be hard, painful, and bloody to beat our powerful creations down. But we must try. We must take risks, give sacrifices, give back, take back.

It's not easy. But when I am scared or intimidated or confused, I listen to The Voice. It is my son's voice. He is saying, *What a continent this must have been.*

Ranting and raving, ranting and raving. Why does wild country bring out this side of me? I'm not sure, but maybe that's why some people, especially politicians and industry leaders, don't want wildernesses like Sylvania set aside, because they remind us of what was, and what could be again. Of what should be. Because places like this could incite revulsion and revolution. Which they should. Would I fight and die for my country? You bet. But I mean country as in countryside, country people, country living.

Ranting and raving. I try—I really do—to just appreciate the wild and rural places, but thoughts like these always intrude. I can't help it. I used to get mad at myself for it; I used to try to cut off that internal bitching and just be a posy-sniffing nature-lover, relishing what is and not worrying about what's coming. But it didn't work. It didn't feel natural. It didn't feel like me. Ed Abbey, again, said it well: "Anger and love. How feel one without the other? Each implies the other." Or maybe Native American singer John Trudell says it best when he warns us not to trust anyone who isn't angry. I don't think I'll bother fighting it anymore. I'll try to keep it all in perspective so I can still appreciate what is, but I'll always remember what was, and what could be. I don't want to let my guard down. I don't want to shirk my responsibility.

We walk on, into the heart of Sylvania. And what a heart

it is. We walk the old, healing road scars that flatten the rolling forest floor and ford the innumerable bogs and marshes and brooks. Without maintenance, the forest recovers on its own—lightning-struck trees fall across the roads, children of maples swarm up the roadsides, ferns and mosses and a remarkable variety of mushrooms coat the road surface. In another generation or two these roads will be barely visible, a memory. In the West, threads of roads such as this have been used to keep vast areas out of the Wilderness Preservation System. Here, though, we see that they don't inhibit the wilderness quality of a place, but rather remind us of the ability of wilderness to heal and return if we don't work hard to maintain our machinations. A good lesson to learn. And here in Sylvania these old roads offer a high-quality, built-in trail system, so the Forest Service doesn't need to get out and make one.

Sarah and I walk side-by-side and talk along this level, green, shrub-covered path. The rain is drenching, soaking, relentless, and we relish it. Our rain gear makes us ready, but the warmth of the rain helps; compared to Colorado's chilly soakers this seems almost a treat, allowing us to enjoy the forest's rain face: The greens are deeper, the scents muskier, the sounds lost in the static of rain hitting leaves falling to leaves. Bogs and swamps and lakes and rain. The North Country is as much water as earth.

At Clark Lake, I stop to take a few casts from the shore, dragging a splashing Jitterbug across the rain-rippled lake surface. Before I am done we can see the change coming. The skies thin, the rain eases, and the mosquitoes begin to fill the void left by the two. We walk on, circling around to the other side of Clark Lake. By the time we reach the portage to Loon Lake, a lake of blue sky approaches from the west like the tide coming in, and we are stripping off our rain gear and replacing it with bug dope.

Where the lodges once stood there are now meadows and

raspberry bushes, and a section of fine beach where a family picnics. We go onward. Our shells are now in our packs, and we have stripped down to t-shirts and rain pants (neither of us thought to wear shorts under them since the sky showed no hint of clearing when we left camp). It is humid as the warm sun lures the moisture back from the ground. The air seems rinsed—the woods are a phosphorescent green and there is no haze whatsoever tinting the view across the lake to the far shore.

A mile or so farther we turn off the old road and follow a thin trail that leads us over a coffee-black bog and to a small, lily pad-ringed pond. Looks like fine, little-visited bass waters. I scramble around the thick shore and finally find a point to climb out on, the piled-and-carefully-organized stick penin-sula of a beaver lodge. Still in use, too; as I step onto the lodge a splashy wake marks the resident's escape to deeper water. From offshore he explodes the lake surface with tail-slaps. I still take a few casts, apologizing to the old boy, but lure noth-ing. After only a few minutes, Sarah and I pull a retreat back to the road, leaving the beaver his little hidden home.

By the time we get back to the old road, the air is dry, the sun is hot, the sky is as dazzling blue as a Rocky Mountain sky. We are overwhelmed by the day. We decide to head back to camp and suck in this new air mass that has moved over us. Through the now-shimmering, now-bird filled, now-photosynthesizing, breathing, humming forest we scoot back, looking forward to a cool swim in Whitefish Lake.

✦ ✦ ✦ ✦

After an afternoon of swimming and reading and nap-ping, night approaches, heralded dramatically by a violent thunderstorm. We sit in our camp chairs on our ground-cloth verandah in front of our canvas condo, and listen to the artillery fire draw near. This is the third major weather pat-tern we have experienced in 12 hours—steady rainfall, to clear and warm sunshine, to wicked thunderstorm. Ah, the

Midwest. And what a thunderstorm it promises to be, I think, as it grumbles, barks, and paws at the earth. We have heard constant pounding for nearly an hour now.

I head down to the canoe (munching raspberries) and paddle out onto the middle of the lake to fill our water bottles. From there I see to the west a leaden mass filling the sky. A huge anvil sweeps off the top and overhead like long hair in a strong wind. It's a big mother. From inside its gut, great flashes erupt every few seconds. Ground-level metallic cumulus clouds flank its thick stalk like destroyers escorting the Bismarck. Like dust clouds rising from its kicking feet. Like I'd better paddle my ass off this lake.

I hustle back, pull the canoe far onto the shore and lean it against a tree. I scramble up the trail to camp, do night chores, pull the bear bag into a tree, stash gear into dry bags, put weight on the little table, brush my teeth. I pass the next hours in the tent, writing to candle-lantern light and peering out through a small tent-screen opening at the light-show lightning backed by echoes of thunder claps rolling over each other. It's a little scary, and a lot exciting.

◆ ◆ ◆ ◆

The next morning offers yet another weather cycle. I awake in the thin light of early morning to the sloppy sloshing of rain falling through leaves. Not very motivating. I lie in the half-dark until I realize that half-dark is all this morning will offer. I decide the day is worth getting up for anyway—we are still in Sylvania, after all—so I crawl from my bag, leaving Sarah to snooze, and realize the morning is as cold as it is wet. "Today was cold, grey, and wet," I will scratch into my journal later. But first, I need a cup of hot tea.

Wrapped in a thick pile jacket and a rain coat, I head out into the world. The first thing I notice is that the audible wetness is not rain, but drainage from trees, a steady dripping like one of those ornate fountains, tier to tier to tier ("...there's a tear in my beer...") to ground. Last night's rain is still working its way to earth. And the earth, I notice, has tried to climb up. Everything—tent, stools, table, dry bags, fishing tackle, trees—is splattered with mud from last night's pounding downpour. Despite the dampening effect of the forest canopy, enough hard rain got through to pummel and splatter the duff. I zip up the neck of my jacket, pull out and fire up the little stove, put on some water, and walk down to the lake.

There I find wind. A stiff, steady blow pushes a lemming-line of whitecaps onto the rocks at my feet. Silver-rimmed steel-grey clouds march eastward in formation, occasionally parting ranks to reveal puddles of blue, then closing again. I think of winter ice floes on Lake Superior. I shiver and think of fall, even though it's only early August. Like the Rockies, fall is never far off up North. This is fall's home; it never packs up and leaves, it just takes a short vacation.

Back to camp, where Sarah and I share a breakfast of tea and oatmeal. The wind blows on, but not long after breakfast the ice-floe clouds give way to open-water sky. It is deep, rich, sweet, blue, clear, crisp, and rinsed clean from the rains. Like a high-altitude sky, a high-latitude sky.

Despite the wind, despite the temperature hanging in the low 50s, we stick to our game plan. Paddle. We make the grueling exit from Whitefish, paddling to the portage, portaging to the paddle, a short paddle to the next portage then a long portage to the next paddle, then the pull through the slot between the lakes. I'm beat by the time we reach Clark Lake, more than an hour and only a mile and a half or so from camp. Next time we get a small canoe, I swear. Maybe one of those plastic ones.

I quit whining enough to notice, again, that Clark Lake is beautiful. Outward it is as blue as the sky; downward it is clear and green, the bottom visible for... how far? Depth is impossible to gauge. I can make out green logs over green boulders on a green silty bottom. The image quivers like a dream. The lake also looks a tad ominous. We're in a wide, half-moon cove here, protected from the wind in the lee of a long point. But to the north, beyond the point where thin trees give way to water, a line of whitecaps and an ink-blue lake-top mark where the wind has a long reach, building up a healthy push. Our goal had been to circle the shoreline of this, one of the biggest lakes in Sylvania, but looking at the wind-ripped lake, we're no longer so sure. So we do the most logical and least courageous thing: We stop for lunch in a sunny, sheltered stand of trees near where we entered the lake, and we ponder our fate.

Lunch only delays the inevitable and doesn't last nearly long enough. We zip on our life jackets and turn to face what lies beyond the point. We pass over calm, smooth water, break the point, then face the waves. The boat rocks sickeningly until we point it into the gusts. We look up the lake— it's miles to the green-banded north shore. We lean forward and take long strokes into the stiff-arm of the wind.

We stay at it, struggling to keep the boat straight, not daring to stop stroking lest we get pushed back onto the point, for twenty minutes or so. We advance only a hundred yards. To circle the lake at this rate, we calculate, would take about 12 days. Two weeks to be safe. And it would demand more calories than we could get from all the Twinkies in Michigan. Executive decision: We turn a graceful and grateful retreat, and ride the swells without a stroke back into the lee of the point in three quick minutes. And from there, our retreat carries us all the way back to camp.

But we are not defeated. Our goal was to circle Clark Lake and we shall persevere. Still cool, clear, and windy, it's not a good day for a big-lake paddle, but it is a perfect Northern day for hiking. With a tip of the hat to the North Wind, we trade our sandals for boots, stuff our day packs with water, sausage, cheese, and peanut mix, and we take to the trail.

Our route is an eight-mile loop, some on old Sylvania Club roads but most on trail. Again we start out along the portage leading away from Whitefish Lake. Less than a mile on this road, though, and the trail we want turns left, into the forest. As we take our first steps on the footpath it immediately feels good to be on a route that rolls with the contours of the land—and it is a rolling, hummocky land. The trail bends around the big trees, takes wide turns avoiding wetlands and brook crossings.

We come upon a big old hemlock, recently killed. Like maybe last night. A twisting tear runs up the length of it, spiraling a turn and a half over forty or fifty feet and exposing a woody gash more than a foot wide. Fresh sap stands like beads of sweat on the exposed grained flesh. On the ground we find pieces and chunks and strips of bark and wood, a few as square and long as lumber, strewn as far as twenty yards away. Sarah points out more debris hanging high overhead in nearby trees. I pick up a football-sized chunk and sniff the rich, pungent, sappy scent. This tree just exploded from within, perhaps its sap flash-boiled, vaporized, dynamiting the tree's cambium veins. Last night's lightning storm *was* close.

A couple of centuries of life, for these old-growth hemlocks can live three or four hundred years, is over here in a quick flash-clap. What a sight it must have been, but I'm glad I missed it. Its life is over, but still this old timer's usefulness—essentialness—here in Sylvania continues; it has just entered a new phase. Over the next several years, this tree will dry and fall, possibly tearing a hole in the forest canopy as it drops one night in a wind storm. The opening will let in sunlight that will create in this shady stand of hemlocks a pocket of shrub and bushes. Maybe a few small white birch, the first trees to cover disturbed areas and that need sunlight to germinate and grow, will find footing. Once the ground in this small opening is again shaded by these new plants and trees, the true old-growth trees of the area, the so-called "climax" trees that create the long-term shade that their own seedlings need to survive, will come back. Maybe that will be another hemlock, or this time a sugar maple or yellow birch.

While that forest-succession story is unfolding, this old dead tree will still be doing a lot of work. As a standing dead it will become a dry snag, a home and food to rodents and insects and the birds that feed on those insects, such as chickadees, woodpeckers, bluebirds, and Wood ducks. After the tree falls, those insects may be meals for scavenging bears, the rodents dinner for coyotes. This fallen tree will also be nourishment for the forest itself, returning nutrients to the soil while feeding seedlings, fungi, bacteria, and orchids, which lack the chlorophyll needed to manufacture food from sunlight. Some orchids can get very old, like the Lady's slipper, which takes nearly twenty years from seedling to first bloom.

Here, like everywhere, old-growth forests are not dead or dying, which was for a long time the root philosophy of industrial-scale logging. And logging, even careful logging of old or dead trees, removes vital elements and biomass from the forest system. Areas of catastrophe, on a small scale like this lightning strike, and on a grander scale such as forest fire and windstorm, events that trigger the process of forest succession, are a key defense mechanism in healthy forests. As areas of the forest are pushed back to earlier stages in succession, a monoculture is avoided and a barrier is erected to the spread of insect blights and disease that might afflict one species.

There is another type of old growth here besides the trees. Sylvania's hydrologic system is ancient, matured, complex, functional, essential, and representative of a unique ecosystem that has largely been destroyed. As we walk beyond the exploded tree and into the forest between the lakes, we are out of sight of the lakes, but not of water. The lakes are not the only water bodies here, and in terms of helping wildlife, the lakes aren't even the best water bodies. Hidden throughout the Sylvania's woods are marshes and swamps, the former shallow, open water bodies rich in organic matter and plant life, and the latter similar but forested with cedar, black spruce, and tamarack.

The most remarkable of Sylvania's water bodies, though, is the most hidden. Bogs are stands of water over which, over a long period of time, plants have formed a solid mat, like a layer of warm, living ice. The plant life that comprises this reservoir's roof is tiny and diverse, yet a bog's mat can be strong enough to support a person's walking, which may cause the trees that have grown through the mat and into the water below to sway eerily with each step. And a person who is unaware of what she has stepped onto—perhaps a muskeg, the Northern term for a completely covered bog—may even break through. Has anyone died this way? I have no idea, but the possibility of death by muskeg-miring or full bog-submersion is enough to wake me up to where I step.

Like other things in the wilderness, we cannot let such fears of danger taint an appreciation of the wonder and value of these things. And bogs and muskegs are wonderful and valuable. Even their deceptive and dangerous covering is important. In times of drought, this natural rooftop protects the water reservoir below from evaporation that in droughts will slowly drain other open-water areas. And in unusually wet times, the absorbing power of the thick plant-matting (so absorbent that Indians used bog mats as diapers) can dampen floods. This natural cover is also, without nutrient-rich soil underfoot, where many of the carnivorous plants grow, including the Pitcher plant, Sundew, and the Bladderwort. And amidst the hidden water and rich plant life, live shrews, voles, lemmings, and other small animals that in turn feed the foxes, owls, coyotes, and—until recently—wolves that inhabit the woods. There's more: In the summer the cool air surrounding these areas offers a respite to deer and their fawns, and at one time to the native Wisconsin elk, exterminated by market and trophy hunters in the mid-19th century. Hunkered down near here are other Sylvania residents: lynx and bobcat, black bear, fishers (also killed off in the 19th century for their fur, but successfully reintroduced in the 1960s), pileated woodpeckers, and bald eagles.

Bogs and swamps and marshes are diverse and valuable, all interlocked with the lakes and forests and wildlife and with each other into an ancient evolved living system, an old-growth water system. But (and you know where this is heading) they are also rare and endangered. As much as 95 percent, some estimates say, of the North Country's marshes and swamps and bogs have been drained and dried. As Wisconsin first tried to compete as a grain-growing state in the late 19th and early 20th centuries, and later as it moved into the regal niche of being "America's Dairyland" (we all need some claim to fame, eh?), wetlands were drained for farms and fields. In the Upper Peninsula of Michigan, massive clearcutting of the forests dried out the soil underneath, and the giant slash piles left behind fueled great, hot, unnatural fires that scorched the land near the turn of the century. Sylvania, with its diverse and moist forest intact, survived the fires thanks to its size.

And today the people flock to see this diverse and now-rare forest, this remnant, this survivor. Like us. Sarah and I walk the eight-mile loop through forest and around Clark Lake that few, judging from the little wear on the trail, probably do in its entirety. But as we get near the north end of Clark Lake, near the park's northern boundary and near the road access, we see more and more people-sign. First campsites flagged by colorful tents and camping gear and canoes pulled onto the rocky shore; then a thicker, more beaten trail as we near the picnic area that you can drive to; then, as we reach the north shore, leaving the Wilderness Area and entering the Sylvania Recreation Area, we enter the picnic area itself—a half-dozen picnic tables lined along a manicured beach complete with a life-guard's chair (deserted), and a football-field size planting of Kentucky bluegrass (why?

Something for the rangers to do, mowing this thing?). Then we cross through a thick stand of thistles, European invaders finding a welcome home in the disturbed, heavily used end of the park; then we cross the paved boat launch and parking lot where ten or fifteen cars and trucks sit empty and a grey-haired couple load a sea kayak onto the roof of their little car; then we enter back into the woods, then the Wilderness Area where we pass a few more campsites with kids playing along the shore. And at last, we enter into the thick woods again and back onto a thinner, lesser-used trail. We pass more hidden swamps, marshes, bogs, and muskegs. We go back into the heart of Sylvania. Back to our camp.

❖ ❖ ❖

Our last night here is a dark, heavy night. The weight of the thick, black forest feels like a small closet, but one that is alive, respiring, transpiring. It is a place we now feel familiar with, at least a little. It is a place we have started a relationship with, a friendship we vow to renew and work on as often as possible.

Blessedly tired (tough paddling, an eight-mile walk, and I carried that aluminum aircraft carrier two and a half miles on various portages), we sit in the darkness around a little campfire and sip coffee, talk, listen. We hear something surprising that it takes us a few minutes to pinpoint. Motor boats. The whines and groans reach us over what must be miles, from the south, from Wisconsin. I suspect the outboard boat engine is the official Wisconsin State Sound, narrowly beating out the snowmobile.

I can take a lot of intrusions in wilderness, but aural pollution is one I cannot stand. Our lives and our world are full of engine noise, and we live with it, but it is, in my mind, one of the worst of the many pollutants we allow ourselves to be subjected to. Let's admit something: Machines are for assholes. But in this machine-society we live in (and don't forget, corporations are machines), we all must be assholes sometime, most of the time, in order to function. So we come out here to the wilderness to remember what it's like to be *human*. That's why machines are not allowed in true Wilderness Areas.

Except Sylvania. One of the compromises made to get the Sylvania Wilderness Area designated was allowing outboard engines on Crooked Lake, one of the largest lakes in the reserve. Bullfeathers, as my grandpa used to say. Now some might call me an extremist for saying that, but to me the extreme is that in a state where more than 99 percent of the lakes allow power boats, they still find it necessary to allow them in a Wilderness Area.

Let it pass, let it pass. Stick to the matter at hand—this fire, this night, my wife. This week has been our first time away together, except in few-hour stints, without our little boy. We miss him severely, feel incomplete without him, but still recognize this time together as something special, magical. "Didn't we used to date?" I ask Sarah. Yep, we did. And in those dating times, those times when we felt most in love, when our love deepened until it sprouted into commitment, we were often in wild country. We got engaged in a canoe on a lake only a few miles from here (she couldn't run away that way). Our relationship was and continues to be forged around discovering, exploring, and adventuring as a team, living with a wilderness attitude both in wilderness and in our day-to-day lives together. Good stuff, methinks. Damned good.

◆ ◆ ◆ ◆

Our last morning, and all the pushy weather has settled. The North has rolled out a red carpet for our departure, I think in my arrogance as I sip coffee and breathe in this remarkable morning. When I die, I want one last, slow, deliberate, absorbed walk through my senses, cutting myself as deeply as I can on the edge of the moment. If I could choose, a morning like this would do fine for that sense-walk. A crisp, clear, summer morning, yet one with the air chilled enough to force me to wear a sweater; a morning soaked in heavy, cool, sweet forest sweat, the musk of soil, the damp breath of the lake. One with the soundtrack of a long, drawn-out, unhurried and confident loon's hoot from the lake. One like this morning.

My wife crawls from the tent, stands, stretches, pauses to look out over the lake. I caress with my eyes her red hair, blue eyes, rounded nose, freckled cheeks, and parted lips taking in her first breaths of the morning and savoring the taste of the same North Country air I breathe.

26

FINAL THOUGHTS: IN BED ONE NIGHT

I LIE IN BED. THE ROOM IS DARK. MY WIFE BREATHES SOFTLY IN SLEEP NEXT TO ME. IN THE NEXT ROOM, MY little boy, only seventeen months old, sleeps silently.

I conjure up the full picture as deeply and vividly as though I were sucking in my last breath: I have a family, a home, a life rich in out-of-door wanderings, a career writing and teaching, and I live all my days on the surreal landscape of mountains and rivers and deserts of the Colorado Plateau.

I am in awe of my good fortune, of the fruits of my labors, of the beneficence of life and land, of the cross-stitch of circumstance that has woven this colorfully patterned fabric of a life I wear.

The intensity of this awareness tugs at me, draws me away, pulls at me with an acceleration I feel in the pit of my stomach. I am caught in a whipping wave of joy—not happiness, not fun, not satisfaction nor complacency, but a deep, thick, rich, soaking joy. My spirit rides this rip-tide of emotion, soaring.

Then I am hit. Knocked off kilter. I tumble and fall and this all-consuming image breaks and shatters like dropped stained glass. And I realize... I am afraid. Terrified. By the feeling? By the intensity? Not really. I am terrified by the possibility—by the *thought* of the possibility—of losing this feeling, of all or anything that created this feeling. I am scared to death to let myself feel it, to enjoy it, to even allow it to happen. I jump off the wave before it breaks.

A vacuum is left in my mind, and it fills with questions: How could I have attached myself so deeply, so completely, so overwhelmingly to something so tenuous, so temporary, so vulnerable as a human being? As a life? As a landscape? The world is so dangerous, and the things we love so fragile and vulnerable.

So much could happen.

Only last summer, as my wife and I sat in a coffee shop in our rural little town, bullets from a drive-by shooting strafed the wall at our backs.

A year and a half ago my then-pregnant wife and I were nearly run down by a college kid relishing the power of his Jeep. My wife ended up in the hospital.

Only a month earlier our house burned to the ground. We were left with literally only the clothes on our backs. I still wonder why I chose to wear suspenders that day.

Two years ago I was told I had skin cancer. Years of running rivers, wandering deserts, and grunting up mountains had taken their toll.

What else? In recent years I have been too close to several lightning strikes. I have come too close to drowning in three different rivers. I have had an avalanche run out at my feet.

More? A river I loved was drowned by a dam. A mountain I worshipped was skinned for a ski area. A valley I had befriended caught the subdivision plague.

Hard thoughts. And the hardest of all? What's ahead?

I think of my son and my guts knot and my mind recoils.

I give him all I can wring from my spirit in affection, dedication, caring, hope, labor, and raw, visceral gut-wrenching love. But how can I do that, how can I risk that, when all those dangers stalk him out there? When the world he is inheriting is so dangerous and in such decay? When the wild landscapes he will need are being eaten alive by my swarming generation?

How can I put all my work and faith and expectations— all I have to offer—into the world I want for him, a wild earth of healthy and wild people, when there is so much and so many out there just waiting, working every bit as hard as I am, to crush, to use, to suck dry all that I love?

I must remember: They have money, technology, machinery, weaponry, the motivation of greed unchecked by ethics or humanity, and a large, blind, and ignorant following; we have intelligence, love, dedication, faith, conscience, joy, simplicity, awareness, and a sense of self encompassing the land, other living things, and the future.

My little boy lies in the next room. How can I not fight and love? I have no choice.

I will love him hard.

I will teach him well.

I will fight for his world.

My wife sleeps next to me. Surrounding me are the most dazzling mountains and rivers and forests and deserts I have ever opened my eyes to.

I have work to do.

ABOUT THE AUTHOR

Ken Wright was born and raised in New England, but matured late in life in the West. He has worked as, among other things, a hydrologist, technical writer, factory worker, bus driver, cook, construction laborer, river guide, park ranger, and reporter. He holds an undergraduate degree in earth sciences and received a master's degree in journalism from the University of Colorado. His writing has been published in a variety of regional and national newspapers and magazines, including *Sierra, Backpacker,* and *High Country News.* He now lives in Durango, Colorado, with his wife and two young children, where he makes his living as a college professor, newspaper columnist, and freelance writer. For real living, though, he is raising a family and spending as much time as possible out-of-doors, and as little time as possible at the keyboard.